Lecture Notes in Bioinformatics
12715

Subseries of Lecture Notes in Computer Science

More information about this subseries at http://www.springer.com/series/5381

Carlos Martín-Vide · Miguel A. Vega-Rodríguez ·
Travis Wheeler (Eds.)

Algorithms for Computational Biology

8th International Conference, AlCoB 2021
Missoula, MT, USA, June 7–11, 2021
Proceedings

 Springer

Editors
Carlos Martín-Vide ⓘ
Rovira i Virgili University
Tarragona, Spain

Miguel A. Vega-Rodríguez ⓘ
University of Extremadura
Cáceres, Spain

Travis Wheeler ⓘ
University of Montana
Missoula, MT, USA

ISSN 0302-9743　　　　　　　ISSN 1611-3349　(electronic)
Lecture Notes in Bioinformatics
ISBN 978-3-030-74431-1　　　ISBN 978-3-030-74432-8　(eBook)
https://doi.org/10.1007/978-3-030-74432-8

LNCS Sublibrary: SL8 – Bioinformatics

This Springer imprint is published by the registered company Springer Nature Switzerland AG
The registered company address is: Gewerbestrasse 11, 6330 Cham, Switzerland

Preface

These proceedings contain the papers that were presented at the 8th International Conference on Algorithms for Computational Biology (AlCoB 2021), held in Missoula, Montana, USA, during June 7–11, 2021. Due to the COVID-19 pandemic, AlCoB 2020 and AlCoB 2021 were merged and held on these dates together. AlCoB 2020 proceedings were published as LNCS/LNBI 12099.

The scope of AlCoB includes topics of either theoretical or applied interest, namely

- Sequence analysis
- Sequence alignment
- Sequence assembly
- Genome rearrangement
- Regulatory motif finding
- Phylogeny reconstruction
- Phylogeny comparison
- Structure prediction
- Compressive genomics
- Proteomics: molecular pathways, interaction networks, mass spectrometry analysis
- Transcriptomics: splicing variants, isoform inference and quantification, differential analysis
- Next-generation sequencing: population genomics, metagenomics, metatranscriptomics, epigenomics
- Genome CD architecture
- Microbiome analysis
- Cancer computational biology
- Systems biology

AlCoB 2021 received 22 submissions. Most papers were reviewed by three Program Committee members. There were also a few external reviewers consulted. After a thorough and vivid discussion phase, the committee decided to accept 12 papers (which represents an acceptance rate of about 55%). The conference program included three invited talks and some poster presentations of work in progress.

The excellent facilities provided by the EasyChair conference management system allowed us to deal with the submissions successfully and handle the preparation of these proceedings in time.

We would like to thank all invited speakers and authors for their contributions, the Program Committee and the external reviewers for their cooperation, and Springer for its very professional publishing work.

March 2021

Carlos Martín-Vide
Miguel A. Vega-Rodríguez
Travis Wheeler

Organization

AlCoB 2021 was organized by the University of Montana, Missoula, USA, and the Institute for Research Development, Training and Advice (IRDTA), Brussels/London, Belgium/UK.

Program Committee

Ludmil Alexandrov	University of California, San Diego, USA
Can Alkan	Bilkent University, Turkey
Mani Arumugam	University of Copenhagen, Denmark
Bonnie Berger	Massachusetts Institute of Technology, USA
Sanchita Bhattacharya	University of California, San Francisco, USA
Chao Cheng	Dartmouth College, USA
Keith Crandall	George Washington University, USA
Colin Dewey	University of Wisconsin-Madison, USA
Ian Dunham	European Bioinformatics Institute, UK
Anton Enright	University of Cambridge, UK
Joe Felsenstein	University of Washington, USA
Pedro G. Ferreira	University of Porto, Portugal
Martin Frith	University of Tokyo, Japan
Debashis Ghosh	University of Colorado, USA
Michael Gribskov	Purdue University, USA
Michael Hawrylycz	Allen Institute for Brain Science, USA
Daniel Huson	University of Tübingen, Germany
Kazutaka Katoh	Osaka University, Japan
Miriam Konkel	Clemson University, USA
Maria-Jesus Martin	European Bioinformatics Institute, UK
Carlos Martín-Vide (Chair)	Rovira i Virgili University, Spain
David H. Mathews	University of Rochester, USA
Aaron McKenna	Dartmouth College, USA
Ryan E. Mills	University of Michigan, USA
Burkhard Morgenstern	University of Göttingen, Germany
Zemin Ning	Wellcome Sanger Institute, UK
Joel S. Parker	University of North Carolina at Chapel Hill, USA
Kay Prüfer	Max Planck Institute for Evolutionary Anthropology, Germany
Knut Reinert	Free University of Berlin, Germany
Walter L. Ruzzo	University of Washington, USA
Russell Schwartz	Carnegie Mellon University, USA
Gordon Smyth	Walter and Eliza Hall Institute of Medical Research, Australia

Peter F. Stadler	Leipzig University, Germany
Alfonso Valencia	Barcelona Supercomputing Center, Spain
Fabio Vandin	University of Padua, Italy
Kai Wang	Children's Hospital of Philadelphia, USA
Matt T. Weirauch	Cincinnati Children's Hospital, USA
Travis Wheeler	University of Montana, USA
Zohar Yakhini	Interdisciplinary Center Herzliya, Israel
Shibu Yooseph	University of Central Florida, USA

Additional Reviewer

Marzieh Eslami Rasekh

Organizing Committee

Sara Morales	IRDTA, Brussels, Belgium
Manuel Parra-Royón	University of Granada, Spain
David Silva (Co-chair)	IRDTA, London, UK
Miguel A. Vega-Rodríguez	University of Extremadura, Cáceres, Spain
Travis Wheeler (Co-chair)	University of Montana, Missoula, USA

Contents

Biological Dynamical Systems
and Other Biological Processes

Learning Molecular Classes from Small Numbers of Positive Examples Using Graph Grammars

Ernst Althaus$^{(\boxtimes)}$ ⓘ, Andreas Hildebrandt$^{(\boxtimes)}$ ⓘ, and Domenico Mosca$^{(\boxtimes)}$ ⓘ

Institute of Computer Science, Johannes Gutenberg University Mainz,
Staudingerweg 9, 55128 Mainz, Germany
{ernst.althaus,andreas.hildebrandt,mosca}@uni-mainz.de

Abstract. We consider the following problem: A researcher identified a small number of molecules with a certain property of interest and now wants to find further molecules sharing this property in a database. This can be described as learning molecular classes from small numbers of positive examples. In this work, we propose a method that is based on learning a graph grammar for the molecular class. We consider the type of graph grammars proposed by Althaus et al. [2], as it can be easily interpreted and allows relatively efficient queries. We identify rules that are frequently encountered in the positive examples and use these to construct a graph grammar. We then classify a molecule as being contained in the class if it matches the computed graph grammar. We analyzed our method on different known groups of molecules defined by structural properties and show that our method achieves low false-negative and low false-positive rates.

Keywords: Graph grammars · Machine learning · Molecular classes

1 Introduction

Molecular structure search is a very important and well-investigated problem in computational biology. There are many approaches to describing substructures and search for them [16]. In [2] we showed how graph grammars can be used to easily describe structures with much more flexibility than it is possible, e.g., in SMARTS. For example, it is easy to describe ring-systems in which the rings can have arbitrary length.

One problem that can appear when searching for molecules is that the structure of the molecules of interest, is unknown. For example, if a researcher identified a few molecules with a common functional property, it may be unclear what the relevant structural property is (if there is a unique structure responsible for this property at all). Our aim is to give computational support to the resulting problem. We assume that the molecules with this property form a molecular class

Supported by the German science foundation (DFG, project number 416768284).

C. Martín-Vide et al. (Eds.): AlCoB 2021, LNBI 12715, pp. 3–15, 2021.
https://doi.org/10.1007/978-3-030-74432-8_1

and we want to learn this class from a small number of positive examples. The researcher can then scan a database for further molecules that are in the same class. In addition, we tried to make the description of the class small and interpretable so that the researcher can check whether the proposed class is reasonable.

We decided to derive an algorithm that learns a graph grammar. A graph grammar is flexible enough to describe even complicated structures in a very compact way and hence leads to interpretable results. Furthermore, the scan of a database is relatively efficient.

We evaluated our approach on different well-known molecular classes, starting with classes of a very simple structure to more complicated classes. For the first, even our simplest approach gives very high true-positive and true-negative rates rates; however, we need some additional ideas to get similarly good results for the more complicated classes.

In an abstract view, our approach can be interpreted as defining a new feature space, namely the set of all graph grammars. From the given examples, we construct a certain subset of these features. Starting from these, we use the following very simple learning algorithm on these features: We use a simple approach to predict the specificity of a feature and select the features that are above some specificity threshold and match the largest number of positive examples (to be precised later). For the molecular classes we investigated, this simple learning approach is sufficient for very good true-positive and true-negative rates. We guess that for even more complicated classes, more advanced learning methods would be helpful.

To show that this new feature space is specific, we compared our approach to simple approaches using random forests, support vector machines, extra tree classifier with only the canonized SMILE as feature. We also compared our approach with the approach of Hiro et al. [6], who participated in the TOX21 competition in 2014. It is a convolutional neural network approach based on SMILES representation of compounds for detecting chemical motifs. The used machine learning approaches serve as a proof-of-concept, demonstrating that our graph grammar-based feature space is advantageous to merely using the canonical SMILEs directly, and that machine learning approaches on small molecules could profit from its use.

The paper is organized as follows. In the next section, we summarize some related work. In Sect. 3, we describe our basic learning approach and an improvement to better predict the specificity of a graph grammar. In Sect. 4, we describe our experiments and finally give a short conclusion.

2   Related Work

In contrast to the typical biopolymers DNA, RNA, and proteins, where similarity search reduces to problems of substring matching, small molecular compounds can feature a much more diverse chemistry [7] and often have non-linear structures with complex ring-systems. In addition, similarity on small molecules can have quite different definitions in different contexts. For many application scenarios, such as lead identification [15], similarity is defined through a similar binding behavior to molecular targets of interest. If the structure of such molecular receptors is known, virtual screening methods such as FlexX [13] or Glide [4] can be employed. In the absence of the receptor structure, pharmacophore models are often derived from sets of active compounds, which are later queried against molecular databases (see, e.g., [5]). Alternatively, molecular similarity is often derived from sets of pre-defined molecular descriptors. These lead to a so-called fingerprint: a binary vector with one bit per property for each molecule. Similarity of molecules is then computed from a comparison of these fingerprints, such as the popular Tanimoto similarity [3, 12]. More recently, deep neural networks have been trained to work on SMILES-representations of molecules [6, 10]. These methods can, in principle, be used to derive similarity classifiers.

The table shows graph grammar rules for the ALK inhibitor class. The left side of the table represents the searched substructure and the right side the resulting node label. The starting label is S. The square brackets represent exactly one label.

Table 1. ALK inhibitor rules

Replacement graph		Label
C ——— F	⇐	[CF]
——— [CF] ——— C ———	⇐	——— [CFC] ———
——— [CFC] ——— C	⇐	——— [CFCC]
[CFCC] ——— C	⇐	[CFCCC]
[CFCCC]	⇐	S

3 Learning the Graph Grammar

In [2] we introduced a simple and intuitive definition for a graph grammar to describe partial structures of a molecule and gave an algorithm to query a database for molecules with this structure. Furthermore, we analyzed for which types of graphs the algorithm is polynomial and gave certain hardness results for other types of graphs. In the following, we will show a simple algorithm to learn a graph grammar from some positive examples.

The basic idea is that the structure that we want to describe is composed of some patterns, where each pattern can again be composed from other patterns, including the pattern itself so that we have recursion. We represent a pattern by a labeled node in a graph that is already constructed. A structure is generated from a single node with a fixed label. We then iteratively replace a single labeled node, called the replacement node, by another graph, called the replacement graph. In order to do so, we have a finite set of rules giving the replacement graphs for a node with a certain label. Furthermore, the rule specifies the endpoints in the replacement graph of the edges adjacent to the replacement node. Hence, we assume that each rule can only be applied if the degree of the replacement node matches the degree specified in the rule. This type of graph grammar can be used to describe a rich set of graph classes. For an example, see Table 1.

We sketch the basic definitions and results obtained in [2]. For details, we refer to paper [2]. Given sets $\mathcal{L}^V$ and $\mathcal{L}^E$ of node- and edge-labels, a multigraph G is a tuple (V, E, N, L_V, L_E), where V is a finite set called nodes, E is a finite set called edges, N maps a set of two (not necessarily disjoint) nodes to each edge, L_V maps a node-label from $\mathcal{L}^V$ to each node and L_E maps an edge-label from $\mathcal{L}^E$ to each edge. A graph rewriting system over $\mathcal{L}^V$ and $\mathcal{L}^E$ is a tuple (S, P) where S is a label from $\mathcal{L}^V$ and P is a finite set of production rules rules of the form $(L, L_1, \ldots, L_d) \rightarrow (G, n_1, \ldots, n_d)$, where $L \in \mathcal{L}^V$, $L_1, \ldots, L_d \in \mathcal{L}^E$, G is a graph and $n_1, \ldots, n_d$ are nodes of G. We call d the degree of the rule and G the replacement graph. Such a rule allows to replace a node with label L whose incident edges have labels $L_1, \ldots, L_d$ by the graph G. The edges incident to the removed node are assigned new endpoints $n_1, \ldots, n_d$ in this order. A more formal definition of a replacement can be found in [2]. In the following, we will sometimes additionally require that the graph G in a replacement rule is connected.

Let $\mathcal{G}(S, P)$ be the set of all graphs that can be obtained by iteratively applying rules starting from the graph consisting of a single node with label S and no edges. Given a graph and a graph rewriting system, the subgraph-matching problem asks whether a subgraph of G is in $\mathcal{G}(S, P)$. This problem is NP-hard even for trees and graphs of bounded pathwidth and bounded degree. It is $W[1]$-hard in the degree for trees. If the graph has bounded degree and a bounded size of each minimal cut, the problem can be solved in polynomial time. For all these results see [2].

3.1 Learning a Graph-Grammar

In this section, we describe our algorithm to learn a graph grammar, i.e., given a set of examples, we compute a graph grammar. This graph grammar naturally defines the class of molecules, i.e., the molecules such that a subgraph is matched by the graph grammar (GG). Our aim is that all examples lie in the class and that the class generalizes to further molecules that are structurally similar.

Given an edge $uv \in E$ of a graph $G = (V, E, N, L_V, L_E)$ the rule associated with uv is the rule in which a node of label $[L(u), L(v)]$ and degree $|\delta(\{u, v\})|$ is replaced by a graph with two nodes labeled $L(u)$ and $L(v)$ that is connected with as many edges as there are in G between u and v and the two nodes having degrees $|\delta(u)|$ and $|\delta(v)|$. Notice that this rule is identical for all parallel edges uv. When we apply the rule at the edge, we obtain a new graph in which the nodes u and v are shrunk into a new node with label $[L(u), L(v)]$. We call the labels that may appear in the input graphs the atomic labels and we call the others the constructed labels. We can iterate this construction for an edge of a graph obtained by applying the rule. Notice that this label is independent of the degrees, i.e., there may be several rules that map a label $[L_1, L_2]$ with a fixed outdegree to a graph consisting of two nodes with labels L_1 and L_2 with a different number of edges between them and the outgoing edges distributed between the two nodes. If the label $[L_1, L2]$ with this outdegree is used within another rule, several subgraphs with the same set of atomic labels may be derivable from it, if the outdegree of the subgraph is the same. The number of constructed rules grows exponentially with the size of the rules.

Notice that we want to find common subgraphs, but the rules we construct force the outdegree of the rules as high as in the input graph, i.e., a common subgraph of two or more input graphs is not found if the outdegrees of the subgraph are different or distributed differently to the two nodes (i.e., $\equiv [L_1, L_2] \rightarrow = L_1 - L_2-$ and $\equiv [L_1, L_2] \rightarrow -L_1 - L_2 =$ are different rules). In order to compensate for this, we do the following.

If there are several rules which map a single label to different graphs as it has different outdegrees and/or the outgoing edges are distributed differently between the two nodes, we build for each subset of these rules, the rule where the two nodes become an outdegree which is the minimum within this subset. The outdegree of the rule is simply the sum of the outdegrees of the two nodes (in the example above, $= [L_1, L_2] \rightarrow -L_1 - L_2-$ would be created). This rule is the most specific that matched the complete subset.

The basic algorithm has two parameters, namely, the maximal size k of a rule that is generated and a minimal fraction f of the learning set that we want to cover from a single starting label. The algorithm is given a set X of graphs and the parameters k and f, and works as follows: For $i = 2 \ldots k$, we enumerate all labels of size i that can be obtained from one label of size $i - 1$ and one atomic label of two endpoints of an edge in a graph in X. We add all graphs that can be obtained by applying these rules to X. For each rule, we store all the initial graphs on which it can be applied.

After this construction phase, we try to cover as many graphs as possible with a small number of rules that are expressive enough; for example, in the best case, the generated rules create exactly the searched graphs or subgraphs of a functional group. For each $\ell \leq k$, we determine the maximal number m_ℓ of graphs that are covered by a single rule of length ℓ. Let ℓ' be the largest number such that $m_{\ell'}/|X| \geq f$ and R be the set of all rules R of size ℓ' that cover at least $m_{\ell'}$ graphs.

Finally, we construct the graph grammar as follows: From the starting label, we allow all the left hand sides of the rules in R to deviate. Furthermore, we recursively add all rules whose left hand sides can be derived from previously added rules.

Table 2. Structural formulae of the investigated classes

Class	Structure
Hydrazone	$R^1\diagdown$ $C\!=\!=\!NN$ $R^2\diagup$
Sulfate	$R\!-\!-\!SO_4$
Cyanate	$R\!-\!-\!OCN$
Isocyanate	$R\!-\!-\!NCO$
Thiocyanate	$R\!-\!-\!SCN$
Isothiocyanate	$R\!-\!-\!NCS$

The first column indicates the respective functional group, the second column shows the molecular structure. The R represents an organic residue. The hydrazone class has two organic residues.

In an abstract view, the algorithm can be viewed as defining a new feature space (i.e., the set of normalized rules of size at most k of a graph grammar) and using a very simple learning algorithm (selecting the rules which are declared specific, with the highest number of positive fits). In the following, we describe two methods that allow us to better predict the specificity of a constructed rule. We try to remove very unspecific rules. On the one hand, we remove all rules of size two consisting only a single Carbon atom and one of C, N, or O and rules derived thereof. Furthermore, we construct all rules up to our size parameter k (specifying the maximal size of a rule that is enumerated). We only keep a given fraction of the rules in R with the highest specificity value, which we define as follows. We give each element of the resulting label of a rule a value depending on its frequency and similarly for each bond order, atom, covalent bond and sum up all these values. More precisely, all rules of label length k were evaluated according to their chemical probability. To evaluate the rules, 22 million molecules[1] from organic chemistry were scanned in advance (see Table 3). Of these, a certain fraction F, i.e. a third parameter, of the best rated rules were selected. The remaining rules of length k were deleted. After the deletion of these unspecific rules, rules exist that were no longer needed because they can not derivate to a deleted label of length k and so they were also removed. Table 1 shows the rules we learned for the anaplastic lymphoma kinase inhibitors (ALKs). Additionally, we try to automatically optimize the

---

[1] http://zinc.docking.org/subsets/all-purchasable.

parameter k as follows. We start with a parameter value of $k = 5$. If the value of k results in too high false-positive rates within the learn set, it is an indication that the generated rules are too unspecific for the investigated class. In this case, k is increased by one in order to generate rules that may further restrict. On the other hand, if the method for calculating the common rules cannot generate any rule of length k, k is reduced by one. Finally, we evaluate our grammar on all known examples. If the true positive rate is below a certain threshold, we try to improve the grammar by discarding fewer rules that are classified to be unspecific.

22 million molecules were investigated, summing up the number of elements, the number of bonds and the number of covalent bonds. The table shows the respective distribution. (The hydrogen atoms listed here were explicitly written in the canonical SMILES string (*)). The table is structured as follows: Columns one to three indicate the distribution of the elements, columns four to six the respective distributions of the covalent bonds, and columns seven to nine the distribution of the different number of bonds.

Table 3. Distribution of the molecules

Element	#	%	Covalent bonds	#	%	Number of bonds	#	%
C	312.374.294	71%	1	347.049.551	73%	1	88.020.220	20%
N	48.487.451	11%	2	124.890.722	26%	2	214.754.914	49%
O	44.619.904	10%	3	948.242	1%	3	117.703.511	27%
H*	16.193.861	3%				4	18.784.046	4%
S	7.825.885	2%				5	35	0%
F	5.235.224	1%						
Cl	3.494.122	1%						
Br	923.976	0%						
I	82.527	0%						
P	25.500	0%						

4   Experiments

To evaluate our approach, we examined the following functional chemical groups (FCGs) and inhibitors: cyanates, isocyanates, thiocyanates, isothiocyanates, hydrazones, sulfates, the group of angiotensin-converting-enzyme inhibitors (ACEs), anaplastic lymphoma kinase inhibitors (ALKs), Bcr-Abl tyrosine-kinase inhibitors (BCRs), β-Lactamase inhibitors (betas), calicheamicins, COX-2 inhibitors (COX-2s), MEK inhibitors (MEKs), Proteasom inhibitors (Proteasoms) and proton-pump inhibitors (PPIs). The cyanates-, hydrazones- and sulfates are selected to investigate differed chemical functional groups. The inhibitor classes are investigated because they are much more complex than the other presented classes. In contrast to the FCGs, which only consist of a fixed small number of atoms, inhibitors consist of many different atoms for each class. Table 2 lists the structural formulae of the simple classes. In Table 2 we excluded the chemical structures of the inhibitors due to their size. The data set consists of 750 molecules of the cyanates, 750 molecules of the isocyanates, 750 molecules of the thiocyanates, 750 molecules of the isothiocyanates, 750 molecules of the hydrazones, 750 molecules of the sulfates, 366 molecules of ACE inhibitors, 22 molecules of ALKs, 76 molecules of BCRs, 92 molecules of betas, 36 molecules of calicheamicins, 177 molecules of COX2s, 34 molecules of MEKs, 41 molecules of Proteasoms, additionally 41 molecules of PPIs and 500 different randomly selected molecules for the neutral class. The molecules were chosen from the PubChem database [8] and then split into learning sets L and a test set T. To create the learning sets L, 20 molecules were randomly selected from the respective functional group, and the rest of the molecules were used for the relevant test set T. After splitting the molecules, the whole test set T consists of $5,885$ molecules and each learning set L of 20 molecules. Thus clearly that $T \cap L = \emptyset$ holds. We set $k = 5$, $f = 0,75$ and $F = 0.25$ in all experiments (we haven't done a careful evaluation of the best possible choice of the parameters). Results for different choices of the functional groups and inhibitors are presented in Table 4. Columns six and seven shows in which class the most numbers of false-positives (FP) occurred. It is generally known that large molecules can have several functional groups at the same time. This may be the reason why so many FP hits were generated from the FCG classes. The calculations were done on a Windows 10 (64-bit) machine with an Intel i7-8550U CPU with 16 GB of DDR4-2400 RAM and an NVIDIA GeForce MX150 GPU.

The first column describes the functional class. The columns TP, TN, #, LL specify the true-positive and true-negative rate, the number of rules used and the maximal label length for the corresponding instances of the graph grammar approach. Columns six and seven shows in which class the most numbers of false positives (FP) occurred. The first six rows show FCGs the remaining ones represent inhibitors and calicheamicins.

4.1   Comparison to Machine Learning on SMILES

We compared our graph grammar approach (GG) with different machine learning appoaches like random forests (RFs), ExtraTreesClassifier (ETCs) and support-vector machines (SVMs). We also compared our GG approach with that of Hirohara et al. [6] (HIRO). HIROs approach achieves good results in the detection of chemical motifs at the TOX21 Challenge 2014 [9]. The approach is based on a convolutional neural network. RFs, ETCs and SVMs are often used if only a small amount of learning data is provided. There are, of course, plenty of other machine learning methods, that we could have used for our comparisons. To calculate the RFs, ETCs and SVMs results, we used TensorFlow 1.15.3 [1] and scikit-learn 0.23.2 [11]. The learning and test sets for the machine learning approaches are the same as for our approach. The only difference is that

Table 4. Results: GG approach

Class	TP	TN	#	LL	FP class	#
Hydrazone	100.00	91.19	1	1	COX-2	102
Sulfate	99.86	99.15	6	3	Cyanate	30
Cyanate	99.86	99.93	2	3	Isothiocyanate	2
Isocyanate	99.73	97.75	4	3	Cyanate	105
Thiocyanate	99.73	99.87	4	3	Isothiocyanate	3
Isothiocyanate	99.73	99.36	4	3	Thiocyanate	27
ACE inhibitors	97.68	93.48	10	5	Thiocyanate	84
ALK inhibitors	100.00	99.51	4	5	Thiocyanate	11
BCR inhibitors	100.00	98.48	6	5	Hydrazone	33
β-lactamase inhibitors	97.22	99.81	11	5	Hydrazone	5
Calicheamicins	100.00	98.24	2	2	Sulfate	2
COX2 inhibitors	100.00	97.34	11	5	β-lactamase	71
MEK inhibitors	92.85	92.72	6	5	Isothiocyanate	134
Proteasom inhibitors	95.24	100.00	4	5	–	–
Proton inhibitors	100.00	99.50	4	5	Hydrazone	19

the learning set L of the machine learning approaches contains all molecules of the learning sets at the same time. Thus, the approaches receive additionally the negative examples. For the preprocessing step, we applied the sequence to sequence encoder (seq2seq) [14] to the canonical SMILES representations to convert the sequence data into numerical sequences of fixed length (450 characters) to improve the results of the machine learning approaches. For the RFs and ETCs approaches, we used the default settings using 10,000 trees of depth 1,500. The SVMs and HIROs approach results were obtained using also the standard settings. The results are presented in Table 5. It is clear to see that machine learning approaches very often do not produce good classification results if they receive a limited amounts of learning data. HIROs approach shows very good results in ALKs, calicheamicins and PPIs where true-positives rates of 100% were achieved but showed also poor performance compared to the GG approach for the BCR inhibitors. In the case of the calicheamicins, all approaches showed very good results.

The table is structured as follows. The first column defines the respective class. The columns two to six show the true-positive rates of the different approaches. The columns seven to eleven indicate the corresponding true negative rates.

Table 5. Comparison of the results: GG, RF, SVM, ETC, HIRO

Class	True positives					True negatives				
	GG	RF	SVM	ETC	HIRO	GG	RF	SVM	ETC	HIRO
Hydrazone	100.00	30.00	21.64	27.95	80.27	91.19	90.77	94.45	92.22	99.39
Sulfate	99.86	63.29	90.55	72.19	93.42	99.15	96.77	82.13	96.11	99.35
Cyanate	99.86	18.52	11.11	18.66	66.99	99.93	90.22	93.96	88.79	90.00
Isocyanate	99.73	36.03	27.53	36.03	81.23	97.75	80.02	86.12	80.78	97.78
Thiocyanate	99.73	16.99	8.90	17.26	78.63	99.88	92.37	96.48	93.19	97.92
Isothiocyanate	99.73	19.04	9.04	20.55	90.55	99.35	90.69	94.50	89.16	99.19
ACE inhibitors	97.68	1.45	1.45	1.73	77.46	93.49	96.03	92.12	96.24	94.75
ALK inhibitors	100.00	50.00	50.00	50.00	100.00	99.51	99.19	99.50	99.26	97.37
BCR inhibitors	100.00	1.79	0.00	3.57	8.22	98.47	99.71	99.95	99.90	99.78
β-lactamase inhibitors	97.22	69.44	27.78	69.44	98.61	99.81	98.62	98.76	99.03	99.27
Calicheamicins	100.00	100.00	100.00	100.00	100.00	98.24	99.13	98.22	99.03	99.82
COX2 inhibitors	100.00	65.61	49.68	63.69	73.89	97.34	99.80	99.85	99.83	99.18
MEK inhibitors	92.85	57.14	35.71	57.14	64.29	92.72	99.60	99.42	99.69	97.76
Proteasom inhibitors	95.24	61.90	80.95	61.90	90.48	100.00	97.33	93.18	97.94	99.73
Proton inhibitors	100.00	95.24	47.62	95.24	100.00	99.50	99.57	99.42	99.39	98.44

4.2   Comparison to Standard Machine Learning Using Graph-Grammar Rules as Features

In this section, graph grammar rules are used to create a specific binary feature vector (FV) for each molecule instead of the SMILES representation as explained above.

For each rule used in the prediction of any of the classes used in the experiments in Sect. 4, we create a binary feature being 1 if the rule matches a molecule. This results in a binary feature vector of length 79 (as we use in total 79 rules in characterization of the 15 classes), which contains all information we used to classify the molecules.

The FVs were created as follows: First, with the algorithm from Sect. 3, an attempt was made to apply the selected rules of the respective class to the entire learn and test set from Sect. 4. If a certain rule could be applied to a certain molecule, the FV was given a 1 at the position of the rules used, otherwise the FV was given a 0. In Sect. 4 for each class a different set of rules was learned, e.g., for each class a different number of bits are used to generate the FV substring (one for each class). In the end, all generated FV substrings for each molecule were concatenated and merged into one and receive the class assignment.

The partitioning of the learning and test is the same as described in Sect. 4. The machine learning models from Sect. 4 were not modified. The results are presented in Table 6. When using FVs from the rules of the GG approaches, it is clearly shown that much better results were produced. The table indicates the superiority of using GG rules as features compared to the canonized SMILE. Very often allocation rates of 100 percent were achieved when using FVs (ALKs, calicheamicins and protons). Only for the BCRs a true-positive value of less than

10 percent was obtained in each case, which nevertheless represents an high increase over the values in Table 5.

Table 6. Results using 79 *FV*s as input.

Class	True positives				True negatives			
	GG	RF	SVM	ETC	GG	RF	SVM	ETC
Hydrazone	100.00	84.79	82.74	84.79	91.19	99.88	99.96	99.88
Sulfate	99.86	99.86	99.86	99.86	99.15	99.28	99.32	99.28
Cyanate	99.86	87.93	80.93	87.65	99.93	99.59	99.96	99.54
Isocyanate	99.73	99.32	99.04	99.32	97.75	99.71	98.41	99.65
Thiocyanate	99.73	96.71	94.11	96.16	99.88	99.92	99.90	99.92
Isothiocyanate	99.73	97.81	96.85	97.81	99.35	99.75	99.96	99.75
ACE inhibitors	97.68	87.57	91.91	87.57	93.49	98.18	97.87	98.18
ALK inhibitors	100.00	100.00	100.00	100.00	99.51	99.35	98.27	98.85
BCR inhibitors	100.00	7.14	3.57	3.57	98.47	99.35	99.56	99.31
β-lactamase inhibitors	97.22	98.61	98.61	98.61	99.81	98.80	99.64	98.80
Calicheamicins	100.00	100.00	100.00	100.00	98.24	99.97	100.00	98.80
COX2 inhibitors	100.00	71.34	100.00	71.34	97.34	99.28	99.13	99.26
MEK inhibitors	92.85	85.71	92.86	85.71	92.72	97.32	99.50	97.32
Proteasom inhibitor	95.24	85.71	90.48	85.71	100.00	99.49	100.00	100.00
Proton inhibitors	100.00	100.00	100.00	100.00	99.50	99.80	100.00	99.80

By using *FV*, the RFs, SVMs and ETCs results were able to achieve almost identical results as those of the GG approaches (see Table 6), in contract to the results using the SMILES representation (i.e. those from Table 5). Here, almost all approaches show the same results. Only the BCRs rates were below average. This shows that if we carefully select rules for the features, i.e., rules that are specific, high assignment rates can be achieved.

5 Conclusion

We showed that we can use a very simple learning approach to successfully classify molecules from a small number of positive examples, if we use the matching of certain graph grammars as features. Furthermore, we showed that we can enumerate these features efficiently enough for the examples considered in the paper.

The table is structured as follows. the first column indicates the chemical group. Columns two to nine show the comparison of the results between the GG approach and the machine learning approaches RFs, SVMs, and ETCs.

In future work, we will further improve upon the efficiency of the learning and classification algorithms. Furthermore, we will use more advanced learning algorithms on the features we defined to even improve the results we obtain. In order to do so, we will need to consider classes of molecules that are harder to describe in terms of their structure.

References

1. Abadi, M., et al.: TensorFlow: large-scale machine learning on heterogeneous systems (2015). https://www.tensorflow.org/, software available from tensorflow.org
2. Althaus, E., Hildebrandt, A., Mosca, D.: Graph rewriting based search for molecular structures: definitions, algorithms, hardness. In: Software Technologies: Applications and Foundations - STAF 2017 Collocated Workshops, Marburg, Germany, 17–21 July 2017, Revised Selected Papers, pp. 43–59 (2017). https://doi.org/10.1007/978-3-319-74730-9_5, https://doi.org/10.1007/978-3-319-74730-9_5
3. Bajusz, D., Rácz, A., Héberger, K.: Why is Tanimoto index an appropriate choice for fingerprint-based similarity calculations? J. Cheminformatics **7**(1), 20 (2015). https://doi.org/10.1186/s13321-015-0069-3
4. Friesner, R.A., et al.: Extra precision glide: docking and scoring incorporating a model of hydrophobic enclosure for protein-ligand complexes. J. Med. Chem. **49**(21), 6177–6196 (2006). https://doi.org/10.1021/jm051256o
5. Gohlke, H., Klebe, G.: Approaches to the description and prediction of the binding affinity of small-molecule ligands to macromolecular receptors. Angewandte Chemie Int. Ed. **41**(15), 2644–2676 (2002). https://doi.org/10.1002/1521-3773, https://onlinelibrary.wiley.com/doi/abs/10.1002/1521-3773
6. Hirohara, M., Saito, Y., Koda, Y., Sato, K., Sakakibara, Y.: Convolutional neural network based on SMILES representation of compounds for detecting chemical motif. BMC Bioinform. **19**(Suppl 19), 526–526 (2018). https://doi.org/10.1186/s12859-018-2523-5, https://www.ncbi.nlm.nih.gov/pubmed/30598075
7. Hoffmann, T., Gastreich, M.: The next level in chemical space navigation: going far beyond enumerable compound libraries. Drug Discov. Today **24**(5), 1148–1156 (2019). https://doi.org/10.1016/j.drudis.2019.02.013, http://www.sciencedirect.com/science/article/pii/S1359644618304471
8. Kim, S., et al.: Pubchem substance and compound databases. Nucleic Acids Res. **44**(D1), D1202–D1213 (2016)
9. National Center for Advancing Translational Sciences (NCATS): Tox21 data challenge 2014 (2014). https://tripod.nih.gov/tox21/challenge/
10. O'Boyle, N., Dalke, A.: DeepSMILES: an adaptation of SMILES for use in machine-learning of chemical structures (2018). https://doi.org/10.26434/chemrxiv.7097960.v1, https://chemrxiv.org/articles/DeepSMILES_An_Adaptation_of_SMILES_for_Use_in_Machine-Learning_of_Chemical_Structures/7097960
11. Pedregosa, F., et al.: Scikit-learn: machine learning in Python. J. Mach. Learn. Res. **12**, 2825–2830 (2011)
12. Rogers, D.J., Tanimoto, T.T.: A computer program for classifying plants. Science **132**(3434), 1115–1118 (1960). https://doi.org/10.1126/science.132.3434.1115, https://science.sciencemag.org/content/132/3434/1115
13. Schellhammer, I., Rarey, M.: FlexX-Scan: fast, structure-based virtual screening. Proteins Structure, Funct. Bioinform. **57**(3), 504–517 (2004). https://doi.org/10.1002/prot.20217, https://onlinelibrary.wiley.com/doi/abs/10.1002/prot.20217

14. Sutskever, I., Vinyals, O., Le, Q.V.: Sequence to sequence learning with neural networks. CoRR abs/1409.3215 (2014). http://arxiv.org/abs/1409.3215
15. Xiang, M., Cao, Y., Fan, W., Chen, L., Mo, Y.: Computer-aided drug design: lead discovery and optimization. Comb. Chem. High Throughput Screening **15**, 328–37 (2012). https://doi.org/10.2174/138620712799361825
16. Šípek, V., Holubová, I., Svoboda, M.: Comparison of approaches for querying chemical compounds. In: Gadepally, V., et al. (eds.) DMAH/Poly -2019. LNCS, vol. 11721, pp. 204–221. Springer, Cham (2019). https://doi.org/10.1007/978-3-030-33752-0_15

Can We Replace Reads by Numeric Signatures? Lyndon Fingerprints as Representations of Sequencing Reads for Machine Learning

Paola Bonizzoni[1]([✉])[ID], Clelia De Felice[3][ID], Alessia Petescia[1][ID], Yuri Pirola[1][ID], Raffaella Rizzi[1][ID], Jens Stoye[2][ID], Rocco Zaccagnino[3][ID], and Rosalba Zizza[3][ID]

[1] Dipartimento di Informatica, Sistemistica e Comunicazione, University of Milano-Bicocca, viale Sarca 336, 20126 Milan, Italy
{paola.bonizzoni,yuri.pirola,raffaella.rizzi}@unimib.it,
a.petescia@campus.unimib.it
[2] Faculty of Technology and Center for Biotechnology, Bielefeld University, Universitätsstr. 25, 33615 Bielefeld, Germany
jens.stoye@uni-bielefeld.de
[3] Dipartimento di Informatica, University of Salerno, via Giovanni Paolo II 132, 84084 Fisciano, Italy
{cdefelice,rzaccagnino,rzizza}@unisa.it

Abstract. Representations of biological sequences facilitating sequence comparison are crucial in several bioinformatics tasks. Recently, the Lyndon factorization has been proved to preserve common factors in overlapping reads [6], thus leading to the idea of using factorizations of sequences to define measures of similarity between reads. In this paper we propose as a signature of sequencing reads the notion of *fingerprint*, i.e., the sequence of lengths of consecutive factors in Lyndon-based factorizations of the reads. Surprisingly, fingerprints of reads are effective in preserving sequence similarities while providing a compact representation of the read, and so, k-mers extracted from a fingerprint, called k-fingers, can be used to capture sequence similarity between reads.

We first provide a probabilistic framework to estimate the behaviour of fingerprints. Then we experimentally evaluate the effectiveness of this representation for machine learning algorithms for classifying biological sequences. In particular, we considered the problem of assigning RNA-Seq reads to the most likely gene from which they were generated. Our results show that fingerprints can provide an effective machine learning interpretable representation, successfully preserving sequence similarity.

Keywords: Sequence analysis · Lyndon factorization · Read representation · Machine learning · Sequence mining

1 Introduction

In the Big Data era, characterized by a massive data growth, mining *sequence* data has attracted a lot of attention, since knowing useful patterns from

C. Martín-Vide et al. (Eds.): AlCoB 2021, LNBI 12715, pp. 16–28, 2021.
https://doi.org/10.1007/978-3-030-74432-8_2

sequences can benefit many applications, such as, event prediction, pattern discovery, time-aware recommendation, DNA detection, and feature embedding [12].

In the specific context of *biological sequences*, finding machine-interpretable representations for sequences that can increase performance of machine learning algorithms, is a challenging task. Indeed, even the most sophisticated algorithms would perform poorly with inappropriate features, while simple methods can potentially perform when fed with appropriate features. Most of the approaches proposed in literature adopt existing methods in Natural Language Processing with the goal of discovering functions encoded within biological sequences [1, 11,16]. As well as the common techniques in Bioinformatics to study sequences, also such methods involve fixed-length overlapping n-grams [18,20].

A main question addressed in this paper is whether there exists a "similarity signature" that may be used to represent a sequencing read and that can be easily detected while reading the read itself. We answer to this question by exploiting the *Lyndon factorizations* for a collection of reads. The Lyndon factorization is one of the most well-known factorizations in combinatorics on words: it is unique for a word and it can be computed in linear time [7,14]. The notion of Lyndon word is not novel in the field of Bioinformatics, since it was used to locate short motifs [8] and more recently it was explored in the development of bijective Burrows-Wheeler Transform [13]. Such a factorization has a main desired property: a read shares a set of consecutive common Lyndon factors with the Lyndon factorization of a superstring of the read itself [4,6]. Surprisingly, in this paper we discover that the length of factors in a Lyndon factorization is enough to define a notion of signature that captures sequence similarity: this is our notion of *fingerprint* of a read. Given a fingerprint f which is a sequence of integer numbers, we extract k-fingers, i.e. k-mers of f. Then collections of k-fingers are used as a main signature to analyze a sample of reads.

To show the effectiveness of such a novel representation approach, we explore their use in the framework of RNA-Seq data classification consisting in assigning each read in a collection to the origin gene. A similar problem for RNA-Seq data analysis in trascriptomics, which is filtering reads by origin genes, has been recently considered in [9] where we refer for the main reference literature. Results of such a preliminary evaluation study, show that the fingerprint can be used successfully to provide a machine-interpretable representation and opens up new perspectives on the possibility of defining a new biological sequence embedding technique. This paper is organized as follows. In Sect. 2, we present the factorizations used to deal with the double-stranded nature of sequencing reads. In Sect. 3 we explore results motivating the use of k-fingers by extending probabilistic results for sequence similarity based on k-mers. Section 4 provides the details about the experiments and related methodology carried out to assess our approach. Finally, Sect. 5 discusses the results and future directions.

2   Lyndon Factorizations and Overlapping Reads

Let Σ be a finite alphabet and let $s = c_1 \cdots c_n$ be a sequence of n characters drawn from Σ; we say that s is a *string* over Σ of length n. The *length* and the character c_i (at position i) will be denoted by $|s|$ and $s[i]$, respectively. The *substring* of s from position i to position j will be denoted by $s[i : j]$. A *prefix* or *suffix* of s are substrings $s[1 : j]$ and $s[i : |s|]$, respectively (denoted also by $s[: j]$ and $s[i :]$). A prefix or a suffix of s is *proper* if $j \neq |s|$ and $i \neq 1$. In the following, we will use the notation $s < v$ ($s > v$) to specify that string s is lexicographically smaller (greater) than string v, and the notation $s \ll v$ to specify that $s < v$ and additionally s is not a proper prefix of v.

Now we introduce the definitions of *factorization* and *fingerprint*, which are the main ingredients we use to capture the overlap between two reads. A *factorization* of a string s is a sequence $F(s) = \langle f_1, f_2, \ldots, f_n \rangle$ of factors, such that $s = f_1 f_2 \ldots f_n$. The *fingerprint* of s with respect to $F(s)$ is the sequence $\mathcal{L}(s) = \langle |f_1|, |f_2|, \ldots, |f_n| \rangle$ of the factor lengths. Given a fingerprint $\mathcal{L}(s) = \langle l_1, l_2, \ldots, l_n \rangle$, a *k-finger* is any subsequence $\langle l_i, l_{i+1}, \ldots, l_{i+k-1} \rangle$ of k consecutive lengths, that is, a k-mer of $\mathcal{L}(s)$. The substring $f_i f_{i+1} \cdots f_{i+k-1}$ will be called as *supporting string* of the k-finger.

In order to capture overlaps between reads, we use Lyndon factorizations, namely composed of *Lyndon Words* [3,15]. A string s is a *Lyndon word* if and only if it is strictly lexicographically smaller than any of its proper suffixes. For example, $s = aabbab$ over alphabet $\{a, b\}$ is a Lyndon word, whereas string $s' = abaabb$ is not a Lyndon word, since the suffix $aabb$ is smaller than s'. The Chen-Fox-Lyndon's Theorem [7] states that any nonempty string s has a unique standard Lyndon factorization $F(s) = \langle f_1, f_2, \ldots, f_n \rangle$ such that $f_1 \geq f_2 \geq \cdots \geq f_n$. Such a Lyndon factorization is called CFL from the authors' names. The Duval algorithm [10] allows to compute CFL in linear time and constant space.

A property of the CFL factorization [6], which is crucial in our framework, is the following.

Conservation Property: let s be a string such that $CFL(s) = \langle f_1, f_2, \ldots, f_n \rangle$ and let x, y be substrings of s such that x and y share a common overlap z where $z = f'_l f_{l+1} \cdots f_t f'_{t+1}$ for some indexes l, t with $1 < l, t < n$. Then, $CFL(x)$ and $CFL(y)$ will share the consecutive factors $f_{l+1}, \ldots f_t$, more precisely $CFL(x) = \langle CFL(x'), f_{l+1}, \ldots f_t, CFL(f'_{t+1}) \rangle$ and $CFL(y) = \langle CFL(f'_l), f_{l+1}, \ldots f_t, CFL(y') \rangle$, where $x = x'z$ and $y = zy'$. Similarly, $\mathcal{L}(x)$ and $\mathcal{L}(y)$ will share the consecutive lengths $|f_{l+1}|, \ldots, |f_t|$. It follows that two overlapping strings s and s' will share consecutive common Lyndon factors in their Lyndon factorizations, while the fingerprints of s and s' will share common k-fingers for a suitable k, where $1 \leq k \leq t$, if s and s' share t common Lyndon factors. In this paper, we propose to use the fingerprint as a signature of common overlapping strings. Another type of Lyndon factorization is based on the notion of Inverse Lyndon word: a string s is an *Inverse Lyndon word* if each proper suffix is strictly smaller than s [5]. Then, $F(s) = \langle f_1, f_2, \ldots, f_n \rangle$ is an *Inverse Lyndon factorization* for a string s, if each factor f_j is an Inverse Lyndon word. Bonizzoni et al. in [5] propose a linear time algorithm to compute a special Inverse Lyndon factorization which is unique for s and is called *Canonical Inverse Lyndon factorization* (referred in the following by ICFL). While uniqueness and linear time computation is guaranteed by both CFL and ICFL, a fundamental property of ICFL is that it splits Lyndon words, thus allowing to further factorize too long Lyndon factors of a Lyndon factorization, if needed. We will call CFL_ICFL the factorization obtained by applying first the Standard Lyndon factorization CFL, and then the Canonical Inverse Lyndon factorization ICFL to factors (in CFL) longer than a given threshold T. In other words, given $CFL(s) = \langle f_1, f_2, \ldots, f_n \rangle$, we obtain $CFL\_ICFL(s)$ by replacing each f_i longer than T with $ICFL(f_i)$. Remarkably, CFL_ICFL has the main advantage of producing many factors, thus enriching the set of k-fingers to use for detecting the overlaps. Observe that, while the conservation property holds for CFL_ICFL, since ICFL splits the same Lyndon factors, it is an open problem to formally prove the property for ICFL, which seems to hold also in this case, as suggested in the preliminary experiments.

Moreover, since we want to take into account the double-stranded nature of the reads, we are interested in signatures that are invariant with respect to a read and its reverse and complement. For this purpose, given any factorization F, we define a *double-stranded version* of F, denoted by F^d, having the following property: given $F^d(s) = \langle f_1, f_2, \ldots, f_n \rangle$ then $F^d(\overline{s}) = \langle \overline{f}_n, \overline{f}_{n-1}, \ldots, \overline{f}_1 \rangle$, where $\overline{f}_i$ and $\overline{s}$ are the reverse and complement of f_i and s, respectively. Observe that (when dealing with a double-stranded factorization) the fingerprint $\mathcal{L}(s)$ is the reverse of the fingerprint $\mathcal{L}(\overline{s})$. As a consequence, the two k-fingers supported by the same genomic region on two opposite overlapping reads, are one the reverse of the other. To overcome this fact, we *normalize* the k-fingers, meaning that, given a k-finger $\langle l_1, l_2, \ldots, l_k \rangle$, we take the lexicographically smallest sequence between $\langle l_1, l_2, \ldots, l_k \rangle$ and its reverse $\langle l_k, l_{k-1}, \ldots, l_1 \rangle$, by considering k-fingers as sequences over the alphabet of the natural numbers. In this way, the fingerprints of two reads extracted from the same locus on two different strands share the same normalized k-fingers. We omit the details for space constraints.

3   Probabilistic Behaviour of Fingerprints

In this section, we explore probabilistic results to be used to estimate how k-fingers capture the similarity of two sequences: we first define and analyze the collision phenomenon by estimating the probability that two sequences share some common k-fingers w.r.t. the corresponding notion of k-mers. In the following, let Σ_s be the alphabet of the lengths of the fingerprint of a sequence w. Clearly a k-finger is a k-mer of a fingerprint. Let us define the notion of k-*finger collision* reflecting the fact that a k-finger can be supported by distinct strings.

Definition 1 (k-finger collision). *Let $x, y \in \Sigma^*$ be two sequences. Let F be a Lyndon-based factorization. Let $f(x)$ and $f(y)$ be the fingerprints for x and y with respect to F, respectively. Let $k(x)$ be a k-finger of $f(x)$ and $k(y)$ be a k-finger of $f(y)$. Let $s_{k(x)}$ and $s_{k(y)}$ be two substrings of x and y supporting $k(x)$ and $k(y)$, respectively. If $k(x) = k(y)$ and $s_{k(x)} \neq s_{k(y)}$, then we say that there exists a* collision *between $s_{k(x)}$ and $s_{k(y)}$.*

Since k-fingers of w are k-mers over the alphabet Σ_s, the collision phenomenon can be studied by exploiting results already obtained in literature in the case of k-mers [2,17]. Observe that the length of the fingerprints can vary for each read considered, and so, to extend those results to our case we will indicate with s the *mean length* of a generic set of reads considered.

Let x and y be two random sequences, both of length n, and x_s, y_s the corresponding fingerprints. Then, the probability P_r that x_s and y_s will share a k-finger by chance and not because of a real similarity can be defined as:

$$P_r = 1 - (1 - s^{-k})^{n-k+1} \tag{1}$$

Observe that since s is far greater than the usual alphabets on which reads are defined (e.g. DNA), using k-fingers drastically reduces the occurrence of random matches differently from the usage of k-mers. However, using k-fingers has an intrinsic and unavoidable probability of collision, which is related to the dimension of the set considered and can be computed as illustrated below and easily taken in consideration to limit the effect of this phenomenon in any formula used. Now we will show how to use P_r to compute the expected number of k-fingers shared by two fingerprints of arbitrary length. To this, we first define the probability that a collision occurs, named *collision probability*, and then the probability that none of the reads are corrupted by any error.

Collision Probability. A factorization can be viewed as a function which randomly maps a sequence to a certain number of integers with a uniform distribution. This means that all the integers have the same probability to be picked up. According to this assumption, the probability of k-finger collision is a generalization of the well-known "birthday problem"[1]. Observe that for each item of a k-finger we have s possible values, and so the space of possible values for the k-uple is s^k. Let suppose to pick a single value. After that, there are $s^{k-1}(s-1)$ remaining possibilities that are unique from the first. Therefore, the probability of randomly generating two k-uple that are unique form each other is $\frac{s^{k-1}(s-1)}{s^k}$ $= \frac{s-1}{s}$. After that, there are $s^{k-1}(s-2)$ remaining possibilities that are unique for the first two, which means that the probability of randomly generating three k-uple that are unique is $\frac{s^{k-1}(s-1)}{s^k} \times \frac{s^{k-1}(s-2)}{s^k} = \frac{s-1}{s} \times \frac{s-2}{s}$, and so on. Based on this argument, we can give the following proposition:

Proposition 1. *[k-finger collision probability] Let s be the mean length of the set of sequences considered and let M be the total number of k-fingers generated. Then, the probability that at least two of them are equal is:*

$$1 - \frac{s^{k-1}(s-1)}{s^k} \times \frac{s^{k-1}(s-2)}{s^k} \times \cdots \times \frac{s^{k-1}(s-M-1)}{s^k} \tag{2}$$

which can be approximated by $P_c = 1 - e^{\frac{-M(M-1)}{2s^k}}$

---

[1] https://preshing.com/20110504/hash-collision-probabilities/.

Detecting Similarity Using P_c and P_r. In order to detect similarity between two sequence using fingerprints it is necessary to be able to correctly distinguish whether a common k-finger is coming from a random match, a collision or a shared region. An effective solution could be calculating a threshold using the *Bernoulli distribution*. Let f_1, f_2, be two fingerprints sharing a number x of k-fingers, and let a false match be a single match generated by a random match or a collision. We define as success the event of having one false positive math (a random matching or a collision), to which the probability $P_c + P_r$ as calculated before corresponds. Respectively, we consider as failure both a matching coming from a common region or the case of not having a match, and as number of the experiments the number of k-fingers of the longest between the two considered fingerprint.

Given a dataset of reads, we can use the Bernoulli formula to compute the minimum number x of k-finger matching needed to say that two reads share a common region. Specifically, first we set a threshold δ such that $\binom{n}{x} p^x \cdot (1 - p)^{n-x} < \delta$. Then we compute the solution x which represent the minimum number x of k-finger matching needed to say that two reads are in overlap. So, let f_1, f_2, be two fingerprints sharing a number x' of k-fingers, if $x' > x$ then we can say that f_1 and f_2 are similar with a probability depending on δ.

4   Methodology and Experiments

In this section we provide the details of the experiments we carried out to assess the effectiveness of fingerprints and k-fingers as machine-interpretable representation for sequencing reads. We assume the reader is familiar with the basic notions of machine learning (see [19] for further details). More precisely, the following question is addressed: can fingerprints and k-fingers be used to assign RNA-Seq reads to the correct origin gene? To this aim, we explore two machine learning approaches to classify RNA-Seq reads; the first one is based on fingerprints and the second one is based on k-fingers. We implemented the following five-step methodology, in relation to the input data used in our experiments:

1. *training data collection*: given a **FASTA** file containing the transcript sequences annotated for 6040 genes from chromosomes 1, 17 and 21 (**havana** and **ensembl_havana**; 3 transcripts per gene on average), we have collected the set of all the 100-long substrings extracted from the transcripts of 100 randomly selected genes (we have considered at most 4 randomly selected transcripts for each gene), obtaining a set $\mathcal{C}$ containing 797407 substrings. For each substring we have also collected the ID of the origin gene;
2. *feature extraction*: we have considered the double-stranded factorization algorithms CFL^d, ICFL^d and $\mathsf{CFL\_ICFL}^d$ (with threshold $\mathsf{T} \in \{10, 20, 30\}$) for a total of 5 factorization algorithms. For each algorithm, we have computed the fingerprints of the strings in $\mathcal{C}$ obtaining 5 datasets (each one to be used as feature dataset in the fingerprint-based approach), and then, for each dataset

of fingerprints, we have extracted all the k-*fingers* for k from 3 to 8 obtaining 30 datasets (to be used as features in the k-finger-based approach); we remark that, in the fingerprint-based (resp. k-finger-based) approach, one fingerprint (resp. one k-finger) represents one *sample* in the feature dataset, and each length in such a fingerprint (resp. k-finger) is a *feature*; we remark that the choice of values for k and T is the result of a series of observations on preliminary experiments, and is not linked to the optimization phase of the machine learning models described below.

3. *labeling*: each fingerprint (or k-finger) in a feature dataset is labeled by the ID of the origin gene in order to have a class for each input gene;

4. *validation and classification*: each feature dataset is split into a *training set* (80% of the samples) and a *testing set* (the remaining 20%); the data have been normalized by using the `minmaxscaler` technique; the k-fold cross-validation was performed to validate different machine learning models; some of the most used classification models have been tested on the testing set with the best parameters found during the previous step;

5. *testing*: in order to simulate reads from the set of our 100 input genes in a reliable setting, we have considered a set with 10 Million RNA-Seq reads simulated with Flux Simulator with different gene-expression levels by considering 9403 human genes from chromosomes 1, 17, and 21; then we have extracted from such set the reads originated from the 100 input genes, thus obtaining 285628 reads; we used the best classification model obtained in the previous step to classify the reads by the gene locus; we note that the set of reads obtained was unbalanced, i.e., 142266 reads simulated from the most expressed gene (ID `ENSG00000132517`), and only 2 reads simulated from the least expressed one (ID `ENSG00000116205`); due the multiclass nature of the problem considered, we first calculated the precision, recall, and f-score values for each of the 100 genes and then the average scores.

The last three steps are detailed in the following two sections with respect to the considered approach (fingerprint-based or k-finger-based). We implemented our methodology in Python by using the *scikit-learn* library[2]. All the input and output files and the Python scripts are available online[3].

4.1   Fingerprint-Based Approach

Validation and Classification. By performing a 5-fold cross-validation using the `gridsearchcv` method, we have obtained the best parameters to train and test the following chosen classification models: *Random Forest* (RF), *Logistic Regression* (LR), and *Multinomial Naive Bayes* (MNB). We remark that the goal of this step was to chose the model having the highest accuracy, precision and recall scores for all classes.

---

[2] https://scikit-learn.org/.
[3] https://github.com/rzaccagnino/DeepShark.

Table 1. Results obtained with RF model for each type of double-stranded factorization

Factorization		Accuracy	Precision	Recall	F-score
CFL^d		0.72	0.72	0.72	0.72
$ICFL^d$		0.85	0.85	0.85	0.85
CFL_ICFL^d	T = 10	0.92	0.93	0.92	0.93
	T = 20	**0.93**	**0.94**	**0.93**	**0.94**
	T = 30	0.92	0.93	0.92	0.92

Table 2. Results obtained on CFL_ICFL^d for T = 20 with MNB, LR, and RF

Model	Accuracy	Precision	Recall	F-score
Multinomial Naive Bayes (MNB)	0.42	0.23	0.30	0.44
Logistic Regression (LR)	0.44	0.43	0.30	0.45
Random Forest (RF)	**0.93**	**0.94**	**0.93**	**0.94**

As a result, we observed that the RF model always outperforms the other models, so in the following we report only its results (Table 1). In general the RF model trained with fingerprints obtained by the factorizations CFL_ICFL^d outperforms the other RF models. Moreover, the best results have been obtained by using the factorization CFL_ICFL^d with T = 20, with accuracy 0.93, precision 0.94, recall 0.93, and f-score 0.94. Furthermore, for all the factorizations CFL_ICFL^d the performance does not vary significantly as the parameter T changes. This could be due to the fact that the longest factors to break were the same, and they were most probably longer than 30 bases. In Table 2 we also report the average results obtained by all the models with CFL_ICFL^d with T = 20.

Testing. Best results were obtained using the RF model with CFL_ICFL^d and $T = 20$. More precisely, we obtained an average weighted precision of 0.85, a recall of 0.42, and an f-score of 0.55. Precision and recall are computed for each investigated gene G as $TP/(TP+FP)$ and $TP/(TP+FN)$ (respectively), where TP is the number of reads simulated from G which are correctly assigned to G, FP is the number of reads simulated from a different gene which are erroneously assigned to G, and FN is the number of reads simulated from G which are erroneously assigned to a different gene.

4.2 k-Finger-Based Approach

Validation and Classification. As observed before, also in this approach the RF model always outperforms the other models and so, we only report its results. Results of RF classification tests are shown in Table 3.

The best results have been obtained by using the factorization CFL_ICFL^d with T = 30 and k = 8, i.e., accuracy 0.94, precision 0.94, recall 0.93 and

Table 3. Best results obtained with RF model on CFL^d, ICFL^d and CFL_ICFL^d, T = 5

Factorization	K value	Accuracy	Precision	Recall	F-score
CFL^d	8	0.90	0.93	0.86	0.89
ICFL^d	5	0.91	0.92	0.87	0.88
CFL_ICFL^{d}30	8	0.94	0.94	0.93	0.94

Table 4. Results obtained on the best factorization CFL_ICFL^d for T = 30 and k = 8

Model	Accuracy	Precision	Recall	F-score
Multinomial Naive Bayes (MNB)	0.42	0.35	0.33	0.46
Logistic Regression (LR)	0.41	0.65	0.71	0.50
Random Forest (RF)	**0.94**	**0.94**	**0.93**	**0.94**

f-score 0.94. To compare it with the other models, in Table 4 we report the average results obtained by all the models with CFL_ICFL^d with T = 30 and k = 8. We further analyzed different choices of parameter k and different factorizations (details omitted for space constraints). We obtained results ranging from 0.38 to 0.94 accuracy, according to the type of factorization and the k value. We can notice that in most cases, with increasing values of k the accuracy improves for any type of factorization. Furthermore, for k values between 3 and 6 the best accuracy is always obtained with ICFL^d, while the best accuracy (0.94) was achieved by CFL_ICFL^d with T = 30 and $k = 8$. We also observed that experiments conducted with k $\leq$ 5 show that CFL_ICFL^d factorizations (for T = 10, 20, 30) always result in lower accuracy compared to other methods.

Testing. To evaluate the effectiveness of the k-finger approach to classify RNA-Seq data by the gene locus, we have defined a *Rule-based read classifier* [19]. During several preliminary test experiments we have defined many *criteria* to deduce the classification of a read by the classification of its k-fingers. By empirical observations two possible criteria have been selected: *Majority* and *Threshold*. According to the *Majority* criterium, a gene G reaches the majority for a given read if at least half of the read k-fingers are classified to G; therefore the read is classified to G. The idea of the *Threshold* criterium, instead, is to use the lowest probability whereby a k-finger was correctly classified to a gene. For classifying a read, *(i)* first this value is subtracted from the classification probabilities of each k-finger extracted from the read (*margins*), *(ii)* then the k-finger reaching the highest margin is selected, and *(iii)* finally the read is classified to the gene for which such a margin has been reached by the selected k-finger.

We tested such criteria in different orders and best results were obtained when combined in the following way: *(1)* if the majority is reached, then the read is classified to the gene which achieves majority, otherwise *(2)* the read is classified to the gene which achieves the highest threshold. In Algorithm 1 we describe the classifier based on such techniques. It takes as input the read to classify (**read**), the k-fingers

---

Algorithm 1: Rule-based read classifier

Input   : read, k_fingers, classifier, genes_thresholds.
Output: ID_gene

1 min_threshold ← min(genes_thresholds);
2 classes ← classify(k_fingers, classifier);
3 best_classes, frequence ← most_frequent(classes);
4 **if** *(size(best_classes) == 1) and frequence >= size(k_fingers/2)* **then**
5 | // Majority criteria;
6 | ID_gene = best_classes[0];
7 **else**
8 | // Threshold criteria;
9 | probability_for_gene ← probability_classify(k_fingers, classifier);
10 | max_index_list ← [];
11 | max_margin_list ← [];
12 | **foreach** *sample ∈ probability_for_gene* **do**
13 | | sample_margin ← [(p - min_threshold) for p ∈ sample];
14 | | max_index ← argmax(sample_margin);
15 | | max_index_list.append(max_index);
16 | | max_margin ← amax(sample_margin);
17 | | max_margin_list.append(max_margin);
18 | max_index ← argmax(max_margin_list);
19 | best_class ← max_index_list[max_index];
20 | ID_gene ← best_class;
21 **return** ID_gene;

---

of the read (k_fingers), a k-finger classifier (classifier), and a list containing, for each gene, the lowest probability to be classified by classifier to such gene (genes_thresholds). As output, it returns the ID of the gene to which read is classified (ID_gene). First, the lowest probability is computed by genes_thresholds (line 1). Then, classifier is used to classify k_fingers and so the most frequent genes (best_classes), and the number of occurrences of such genes are extracted by the results of the k-fingers classification (line 2). If best_classes contains only one gene and such a gene reached the majority then it was returned (line 6). Otherwise, for each k-finger sample in k_fingers, given the probabilities to be classified to each gene (line 9), the algorithm first computes the classification margin for each gene (line 13) and then selects the gene for which the highest margin has been reached by sample (lines 15–17). After having computed such a value for each k-finger sample in k_fingers the algorithm returns the first of such genes (line 20).

We used the k-finger classifiers trained in the previous step to implement the rule-based read classifier defined in Algorithm 1. Due to the multiclass nature of the considered problem, we first calculated the precision, recall, and f-score values for each of the 100 genes and then the average scores. Unlike the results obtained during the training step, in which the RF model with CFL_ICFL^d

T = 30 and k = 8 achieved the best scores respect to the k-finger classification, the results obtained for the read classification show that the best scores were achieved by using the RF model with ICFL^d and k = 5. This is because, by reducing the k value, we are able to consider more local parts of each read, obtaining more classifications which make the rule-based read classifier more robust respect to the presence of errors in the read. As first result, we obtained average weighted precision of 0.91, recall 0.77 and f-score 0.82. Given the substantial difference with respect to the scores obtained with the approach of Sect. 4.1, we have decided to perform a deeper analysis of rule-based read classifier performance. We have observed that the precision value was directly proportional to the support value, which is the number of reads assigned to the given gene, with average precision value from 0.99 for the most expressed gene (ID ENSG00000132517) to 0.001 for the less expressed (ID ENSG00000116205). However, the average recall was always greater than 0.62, and in general very high for the less expressed genes (about 1). This means that the classification of the few samples related to such genes is almost always exact, and so the low precision is attributable to the misclassification of the reads related to the most expressed genes. To assess our hypothesis we repeated the same experiment on several balanced subsets of the 285628 reads: *(1)* the subset of reads related to the 31 most expressed genes which have at least 1000 samples (1000 reads × 31 genes), *(2)* the subset of reads related to the 15 genes which have at least 14 samples and less of 100 reads (14 reads × 15 genes), *(3)* 4 reads for each of the most expressed 98 genes (4 reads × 98 genes), and finally *(4)* 2 reads for each of the 100 genes (2 read × 100 genes). As a result, we obtained arithmetic average precision of 0.90, recall 0.77, and f-score 0.77, that by construction corresponds also to the weighted average scores. This result confirms our hypothesis and so we can consider the average weighted precision of 0.91, recall 0.77, and f-score 0.82 as representative results of classifier performance.

5   Discussion

In this paper we investigate the notion of fingerprint as novel signature for representing sequencing reads. While this notion could be used also in a combinatorial setting for comparing biological sequences, being a numeric representation of sequences, we decided to explore its use in a machine learning approach to classify sequencing reads. We experimented different notions of Lyndon-based factorizations, i.e. CFL^d, ICFL^d and CFL_ICFL^d to evaluate the best representation. We assess its potentiality in assigning RNA-Seq reads to their origin gene by carrying out several read classification experiments. The training steps show that both notions in our method allow to achieve comparable high scores. Nevertheless, differences among the use of the two notions emerged during the testing step, where the performances were tested on a simulated RNA-Seq set. In this case, the presence of errors in the reads and the different level of gene expression, leading to an unbalanced number of reads per gene, produce differences in the performance when using fingerprints versus k-fingers of the read.

This significant difference between the results in terms of recall and F-score can be explained with the presence of errors in the reads. Since the fingerprint method considers the whole read, this is sensitive to the presence of errors. Conversely, dividing the reads in k-fingers enables to obtain a more robust classification, allowing to better isolate the corrupted portion of reads, and to identify the correct class with the conjunct application of the rule-based classifier. In conclusion, the results prove the effectiveness of the notions of fingerprint and k-finger as machine learning interpretable representations. This work still has many open questions that need to be addressed, including how fingerprints perform compared to other machine learning representations of biological sequences [1,11] and how the probabilistic analysis presented in the paper can reflect on real sequencing data. Finally, could be fingerprints used in a combinatorial approach to detect the overlap of sequencing reads?

Acknowledgments. This project has received funding from the European Union's Horizon 2020 research and innovation programme under the Marie Sklodowska-Curie grant agreement number [872539].

References

1. Asgari, E., Mofrad, M.R.: Continuous distributed representation of biological sequences for deep proteomics and genomics. PLoS ONE **10**(11), e0141287 (2015)
2. Berlin, K., Koren, S., Chin, C.S., Drake, J.P., Landolin, J.M., Phillippy, A.M.: Assembling large genomes with single-molecule sequencing and locality-sensitive hashing. Nature Biotechnol. **33**(6), 623–630 (2015)
3. Berstel, J., Perrin, D.: The origins of combinatorics on words. Eur. J. Comb. **28**(3), 996–1022 (2007)
4. Bonizzoni, P., De Felice, C., Zaccagnino, R., Zizza, R.: Lyndon words versus inverse Lyndon words: queries on suffixes and bordered words. In: Leporati, A., Martín-Vide, C., Shapira, D., Zandron, C. (eds.) LATA 2020. LNCS, vol. 12038, pp. 385–396. Springer, Cham (2020). https://doi.org/10.1007/978-3-030-40608-0_27
5. Bonizzoni, P., De Felice, C., Zaccagnino, R., Zizza, R.: Inverse Lyndon words and inverse Lyndon factorizations of words. Adv. App. Math. **101**, 281–319 (2018)
6. Bonizzoni, P., De Felice, C., Zaccagnino, R., Zizza, R.: On the longest common prefix of suffixes in an inverse Lyndon factorization and other properties. Theor. Comput. Sci. **862**, 24–41 (2021)
7. Chen, K.T., Fox, R.H., Lyndon, R.C.: Free differential calculus, IV. the quotient groups of the lower central series. Ann. Math. **68**(1), 81–95 (1958)
8. Delgrange, O., Rivals, E.: STAR: an algorithm to search for tandem approximate repeats. Bioinformatics **20**(16), 2812–2820 (2004)
9. Denti, L., et al.: Shark: fishing relevant reads in an RNA-Seq sample. Bioinformatics (2021)
10. Duval, J.P.: Factorizing words over an ordered alphabet. J. Algorithms **4**(4), 363–381 (1983)
11. Kimothi, D., Soni, A., Biyani, P., Hogan, J.M.: Distributed representations for biological sequence analysis. arXiv preprint arXiv:1608.05949 (2016)
12. Kumar, P., Krishna, P.R., Raju, S.B.: Pattern Discovery Using Sequence Data Mining: Applications and Studies. IGI Publishing, United States (2011)

13. Köppl, D., Hashimoto, D., Hendrian, D., Shinohara, A.: In-Place bijective Burrows-Wheeler Transforms. In: Combinatorial Pattern Matching (2020)
14. Lothaire, M.: Combinatorics on Words. Cambridge University Press, Cambridge (1967)
15. Lyndon, R.C.: On burnside's problem. Trans. Am. Math. Soc. **77**(2), 202–215 (1954)
16. Motomura, K., Fujita, T., Tsutsumi, M., Kikuzato, S., Nakamura, M., Otaki, J.M.: Word decoding of protein amino acid sequences with availability analysis: a linguistic approach. PLoS ONE **7**(11), e50039 (2012)
17. Ondov, B.D., et al.: Mash: fast genome and metagenome distance estimation using minhash. Genome Biol. **17**(1), 132 (2016)
18. Srinivasan, S.M., Vural, S., King, B.R., Guda, C.: Mining for class-specific motifs in protein sequence classification. BMC Bioinform. **14**(1), 96 (2013)
19. Tan, P.N., Steinbach, M., Kumar, V.: Introduction to Data Mining. Pearson Education India (2016)
20. Vries, J.K., Liu, X.: Subfamily specific conservation profiles for proteins based on n-gram patterns. BMC Bioinform. **9**(1), 72 (2008)

Exploiting Variable Sparsity in Computing Equilibria of Biological Dynamical Systems by Triangular Decomposition

Wenwen Ju and Chenqi Mou[✉]

LMIB–School of Mathematical Sciences, Beihang University, Beijing 100191, China
{juwenwen,chenqi.mou}@buaa.edu.cn

Abstract. Biological systems modeled as dynamical systems can be large in the number of variables and sparse in the interrelationship between the variables. In this paper we exploit the variable sparsity of biological dynamical systems in computing their equilibria by using sparse triangular decomposition. The variable sparsity of a biological dynamical system is characterized via the associated graph constructed from the polynomial set in the system. To make use of sparse triangular decomposition which has been proven to maintain the variable sparsity when a perfect elimination ordering of a chordal associated graph is used, we first study the influence of chordal completion on the variable sparsity for a large number of biological dynamical systems. Then for those systems which are both large and sparse, we compare the computational performances of sparse triangular decomposition versus ordinary one with experiments. The experimental results verify the efficiency gains in sparse triangular decomposition exploiting the variable sparsity.

Keywords: Systems biology · Biological dynamical system · Equilibria · Variable sparsity · Triangular decomposition · Chordal completion

1 Introduction

Systems biology studies the behaviors of and interactions between different components of complex biological systems from the computational and mathematical viewpoints. Dynamical systems which describe the behaviors and interactions with time dependence, usually in the form of differential equations, are a common mathematical model for many biological problems like chemical reaction networks and signal pathways [1,8,9]. Due to the infeasibility to find the analytical solutions of biological dynamical systems in most cases, qualitative study is the mainstream with respect to their behaviors like stability [5,11,23,29], bifurcations

This work was partially supported by the National Natural Science Foundation of China (NSFC 11971050) and Beijing Natural Science Foundation (Z180005).

[24, 28], limit cycles, and chaos [24]. Because of their good properties like being free of errors, symbolic tools like Gröbner basis [14, 15], triangular set [18], quantifier elimination [7, 13], and real solution classification [29] have been successfully applied to study the dynamical behaviors of differential biological systems.

Biological dynamical systems are usually large in scale, with tens of or even hundreds of nodes in the corresponding network, and sparse in relationship, in the sense that only a couple of nodes are interrelated in the network. The former property brings about computational difficulties to symbolic methods listed above, while the latter sheds light on the use of specialized sparse algorithms for full utilization of the inherent sparse structure of the systems. This paper focuses on exploiting the variable sparsity of biological dynamical systems in the settings of computing their equilibria with the symbolic method triangular decomposition.

The sparsity of biological systems has been studied by using graphical tools like directed graphs and weighted bipartite graphs [10] and strongly connected components [20]. In this paper we use chordal graphs to exploit the variable sparsity appearing in the defining polynomial sets of the biological dynamical systems. Chordal graphs, also called triangulated graphs [26], have unique structural properties so they are widely used in various scientific and engineering problems, for example in discrete and continuous optimization [4] and graphical models [16]. In particular, chordal graphs play a central role in the study of sparse Gaussian elimination for solving large and sparse linear systems by rearranging the coefficient matrices via the perfect elimination orderings of the chordal associated graphs [27]. This idea is then generalized to triangular decomposition for solving polynomial systems: several typical algorithms for triangular decomposition are proved to preserve the chordal structure and an algorithmic framework for sparse triangular decomposition is proposed to make full use of the variable sparsity of the polynomial systems [6, 21, 22].

In this paper we apply the sparse triangular decomposition to compute the equilibria of a large number of biological dynamical systems (141 in total). The main contributions of this paper include: (1) We characterize the variable sparsity of the polynomial sets appearing in the biological dynamical systems with the chordal completions of the associated graphs of the polynomial sets; (2) We exploit such variable sparsity for computing the equilibria of the biological dynamical systems by using sparse triangular decomposition; (3) Our experimental results provide practical information on suitable algorithms for chordal completion and the performance gains of the sparse algorithm for triangular decomposition versus the ordinary one in the settings of computing the equilibria.

This paper is organized as follows. Section 2 provides the background knowledge of equilibria of biological dynamical systems, polynomial sets and their associated graphs, and sparse triangular decomposition. Then in Sect. 3 the variable sparsity of the biological dynamical systems before and after chordal completion is studied and reported for 141 systems. In Sect. 4, for those sparse and large systems, we apply sparse triangular decomposition to compute the equilibria and compare its computational efficiency with ordinary triangular decomposition. We conclude this paper with remarks in Sect. 5.

2   Preliminaries

2.1   Autonomous Biological Dynamical Systems and Their Equilibria

Many mathematical models, as in the databases BioModels[1] [17] and ODEbase[2], of biological systems such as chemical reaction networks and signal pathways are in the following form of autonomous dynamical systems

$$\begin{cases} \frac{dx_1}{dt} = \frac{P_1(x_1,\ldots,x_n)}{Q_1(x_1,\ldots,x_n)}, \\ \quad\vdots \\ \frac{dx_n}{dt} = \frac{P_n(x_1,\ldots,x_n)}{Q_n(x_1,\ldots,x_n)}. \end{cases} \tag{1}$$

In system (1) above, $x_1,\ldots,x_n$ represent the amounts of n reactants which change over time and are considered variables in time t, and their change rates with respect to t, namely $\frac{dx_i}{dt}$ for $i = 1,\ldots,n$, are represented by rational functions in $x_1,\ldots,x_n$ in the right hands (being rational means that both P_i and Q_i in the fraction are polynomials).

To further formulate system (1), we consider the multivariate polynomial ring $\mathbb{R}[x_1,\ldots,x_n]$ over the real number field $\mathbb{R}$ in the variables $x_1,\ldots,x_n$, written as $\mathbb{R}[x]$ for simplicity with $x = x_1,\ldots,x_n$. Then P_i and Q_i are polynomials in $\mathbb{R}[x]$ for $i = 1,\ldots,n$. Furthermore, a point $\overline{x} = (\overline{x}_1,\ldots,\overline{x}_n) \in \mathbb{R}^n$ is said to be a *equilibrium* or *steady state* of system (1) if $P_1(\overline{x}) = \cdots = P_n(\overline{x}) = 0$ and $Q_1(\overline{x}) \neq 0, \cdots, Q_n(\overline{x}) \neq 0$. It is easy to see that at an equilibrium $\overline{x}$, the right hands of (1) are all equal to 0, meaning that none of $x_1,\ldots,x_n$ changes at this state and thus the system is steady.

In most cases the solutions of dynamical systems like (1) cannot be expressed analytically. Therefore, the behaviors like stability and bifurcations of the systems are usually studied qualitatively. As the very start of such a qualitative analysis, all the equilibria of the systems need to be computed, preferably symbolically (meaning without any errors) so that the subsequent stability analysis does not suffer numerical problems incurred by the errors. To compute all the equilibria of system (1), we need to find all the solutions of the polynomial system $P_1(x) = \cdots = P_n(x) = 0, Q_1(x) \neq 0, \cdots, Q_n(x) \neq 0$ derived from the right hand of (1).

2.2   Polynomial Sets and Associated Graphs

In this subsection and the one to follow, for universality we consider a general field $\mathbb{K}$ instead of $\mathbb{R}$. Let $F \in \mathbb{K}[x]$ be a polynomial and $\mathcal{F} \subset \mathbb{K}[x]$ be a polynomial set. Denoted by $\mathrm{supp}(F)$ the set of variables in $\{x_1,\ldots,x_n\}$ which effectively appear in F and $\mathrm{supp}(\mathcal{F}) := \bigcup_{F \in \mathcal{F}} \mathrm{supp}(F)$. The *associated graph* of $\mathcal{F}$, denoted by $G(\mathcal{F})$, is an undirected graph (V, E) with the vertex set $V = \mathrm{supp}(\mathcal{F})$ and the

---

[1] https://www.ebi.ac.uk/biomodels/.
[2] http://odebase.cs.uni-bonn.de/.

edge set $E = \{(x_i, x_j) : 1 \leq i \neq j \leq n$ and $\exists F \in \mathcal{F}$ such that $x_i, x_j \in \text{supp}(F)\}$. As one can see, the associated graph reflects how the variables are interconnected with each other in the polynomial set.

Definition 1. Let $G = (V, E)$ be a graph with $V = \{x_1, x_2, \ldots, x_n\}$. Then a vertex ordering $x_{i_1} < x_{i_2} < \cdots < x_{i_n}$ is called a *perfect elimination ordering* of G if for each $j = i_1, \ldots, i_n$, the restriction of G on the set $\{x_j\} \cup \{x_k : x_k < x_j$ and $(x_k, x_j) \in E\}$ is a clique. A graph is said to be *chordal* if it has a perfect elimination ordering.

For an arbitrary graph G, another graph G' is called a *chordal completion* of G if G' is chordal with G as its subgraph. When there is no ambiguity, the process to compute a chordal completion is also called chordal completion. There are effective algorithms to determine whether a given graph is chordal, such as the Lex-BFS (lexicographic breadth first search) algorithm [27] and MCS (maximum cardinality search) algorithm [3], and there are also algorithms for constructing a minimal chordal completion of a graph (discussed in more details later in Sect. 3.2) [12].

Definition 2. Let $\mathcal{F} \subset K[x]$ be a polynomial set and $G(\mathcal{F}) = (V, E)$ be its associated graph. Then the *variable sparsity* of $\mathcal{F}$ is $s_v(\mathcal{F}) := |E|/\binom{2}{|V|}$, where the denominator is the number of edges of a complete graph composed of $|V|$ vertices.

The variable sparsity is naturally defined via the associated graph which presents how the variables are connected in the polynomial set. The smaller $s_v(\mathcal{F})$ is, $\mathcal{F}$ is sparser with respect to its variables.

2.3   Triangular Sets and Sparse Triangular Decomposition

In the polynomial ring $\mathbb{K}[x]$, let a variable ordering be fixed. For an arbitrary polynomial $F \in \mathbb{K}[x]$, the greatest variable appearing in F is called its *leading variable*, denoted by $\text{lv}(F)$, and the leading coefficient of F viewed as a univariate polynomial in $\text{lv}(F)$ is called its *initial*, denoted by $\text{ini}(F)$. For a polynomial set $\mathcal{F} \subset \mathbb{K}[x]$, the set of common zeros of $\mathcal{F}$ in $\overline{\mathbb{K}}^n$ is denoted by $\mathsf{Z}(\mathcal{F})$.

Definition 3. An ordered set of non-constant polynomials $\mathcal{T} = [T_1, \ldots, T_r] \subset \mathbb{K}[x]$ is called a *triangular set* if $\text{lv}(T_1) < \cdots < \text{lv}(T_r)$. Let $\mathcal{F} \subset \mathbb{K}[x]$ be a polynomial set. Then a finite number of triangular sets $\mathcal{T}_1, \ldots, \mathcal{T}_s \subset \mathbb{K}[x]$ are called a *triangular decomposition* of $\mathcal{F}$ if the zero relationship $\mathsf{Z}(\mathcal{F}) = \bigcup_{i=1}^s \mathsf{Z}(\mathcal{T}_i) \setminus \mathsf{Z}(I_i)$ holds, where $I_i = \prod_{T \in \mathcal{T}_i} \text{ini}(T)$.

The triangular shape of the polynomials in a triangular set with respect to their greatest variables makes it easy to solve an equation set defined by a triangular set via repeatedly solving a univariate polynomial and substitution into the next polynomial, in a similar way to what is done in solving a linear equation set with a coefficient matrix in echelon form. The process of computing a triangular decomposition is also called triangular decomposition, if no ambiguity occurs. Triangular decomposition is a polynomial generalization of Gaussian elimination and is one fundamental tool for solving polynomial systems symbolically. There are a few kinds of triangular sets, like the regular, simple, normal, and irreducible ones, etc., according to different defining properties of the polynomials within [2,31]. In this paper, mainly in Sect. 4, we only use the regular triangular set, whose zero set is guaranteed to be non-empty, and the decomposition into regular triangular sets is called regular decomposition [30].

By making use of the chordal structure, theoretical properties of triangular decomposition in top-down style have been analyzed and it has been proved that algorithms for triangular decomposition in top-down style "preserve" the chordality, namely the associated graphs of all polynomial sets in the process of such triangular decomposition are subgraphs of the chordal associated graph of the input polynomial set. This means that the variable sparsity of the input chordal associated graph (a corresponding chordal completion in case this associated graph is not chordal) imposes an upper bound for that of all the polynomial sets in the whole process of triangular decomposition.

Based on these theoretical results, a sparse version of triangular decomposition in top-down style is proposed in [22] to fully use the variable sparsity and chordality. The proposed sparse triangular decomposition is shown to be efficient experimentally when the polynomial systems to solve are sparse and becomes more efficient when the systems become sparser. In this paper we mainly use this sparse algorithm to exploit the variable sparsity when computing the equilibria of biological dynamical systems as in (1). The readers are referred to Algorithm 4 in [22] for the details of this algorithm.

3   Variable Sparsity in Biological Dynamical Systems

In this section, we report the results of our analysis on the variable sparsity for 141 systems from the ODEbase database, before and after chordal completion of the associated graphs of the input polynomial sets.

3.1   Original Variable Sparsity

We first chose 141 autonomous biological dynamical systems in the form of polynomials instead of rational functions over the rational number field $\mathbb{Q}$ to extract the polynomial sets as the right hands of these systems. Then for each polynomial set, we test whether its associated graph is chordal by using the Lex-BFS algorithm and compute its variable sparsity according to Definition 2. The results are summarized in Table 1 for the polynomial set of less than or equal to 30 variables and in Table 2 for those of more than 30 variables. In these two tables, column "Model" records the ID "xxx" of the considered model whose ID label is BIOMD0000000xxx in the BioModels database, column "n" records the number of variables, column "Chor" records whether the associated graph is chordal ("1" stands for yes, while "0" for no), and column "s_v" records the variable sparsity.

As can be seen from these two tables, 91 out of 141 systems (around 65%) have chordal associated graphs. Furthermore, the 117 relatively small systems of ≤ 30 variables tend to be denser with respect to the variables than the 24 larger systems with more than 30 variables, 22 of which (around 92%) are of variable sparsity ≤ 0.4. Based on these analyses, we focus on the 22 large systems which are sparse and thus suitable for sparse triangular decomposition.

Table 1. Variable sparsity of polynomial sets of ≤ 30 variables

Model	n	Chor	s_v	Model	n	Chor	s_v	Model	n	Chor	s_v	Model	n	Chor	s_v	Model	n	Chor	s_v
002	12	1	1.000	103	17	1	0.993	271	6	1	0.400	389	22	0	0.338	609	5	1	0.500
009	26	1	0.298	108	9	1	0.639	272	6	1	0.400	405	6	1	0.667	619	10	0	0.333
011	22	1	0.420	123	14	0	0.780	282	6	1	0.200	413	5	1	1.000	629	5	1	0.500
026	11	1	0.909	147	24	0	0.630	283	4	1	0.500	430	27	0	0.613	630	6	1	0.400
028	16	0	0.867	150	4	1	1.000	289	5	1	0.600	431	27	0	0.604	637	12	1	1.000
030	18	0	0.863	156	3	1	1.000	292	6	1	0.067	447	13	1	0.667	638	21	1	1.000
035	10	1	0.622	159	3	1	1.000	305	9	1	0.028	459	4	1	0.500	646	18	1	0.255
038	17	0	0.382	160	25	0	0.163	306	5	1	0.100	460	4	1	0.500	647	11	1	0.491
040	5	1	0.300	163	16	0	0.467	307	5	1	0.100	475	23	0	0.364	651	29	0	0.165
046	16	0	0.450	178	6	1	0.200	309	4	1	0.000	483	8	1	0.393	661	7	1	1.000
050	14	0	0.253	197	7	1	0.381	310	4	1	0.000	484	2	1	0.000	663	3	1	1.000
052	11	1	0.200	198	12	1	0.152	311	4	1	0.000	485	2	1	0.000	676	15	0	0.371
057	6	1	0.667	199	15	0	0.095	314	12	1	0.288	486	2	1	1.000	687	15	1	0.229
060	4	1	1.000	200	22	0	0.688	315	19	1	0.731	487	6	1	0.733	688	16	1	0.250
069	10	1	0.756	226	24	0	0.181	321	3	1	0.667	500	14	1	0.659	692	8	0	0.857
072	7	1	0.619	229	7	0	0.571	357	9	1	0.778	519	3	1	1.000	707	5	1	0.800
080	10	1	0.711	230	26	0	0.535	359	9	1	1.000	530	10	1	0.711	708	5	1	0.500
082	10	1	0.711	233	4	1	0.167	360	9	1	0.972	539	6	1	0.800	709	5	1	0.500
085	17	0	0.978	243	23	1	0.225	361	8	1	0.929	546	7	1	0.143	710	7	0	0.667
086	17	1	1.000	257	11	1	0.491	363	4	1	0.500	552	3	1	0.333	716	4	1	0.667
091	16	1	0.442	259	17	1	0.478	364	13	1	0.846	553	3	1	0.333	726	8	1	0.464
092	4	1	0.500	260	17	1	0.478	365	30	0	0.768	569	21	1	1.000				
099	7	0	0.571	261	17	1	0.478	383	6	1	0.067	581	28	0	0.175				
102	13	1	0.923	267	4	1	0.333	384	6	1	0.067	599	30	0	0.501				

3.2   Variable Sparsity After Chordal Completion

As shown in Table 2, the associated graphs of 22 out of the 24 polynomial sets of more than 30 variables are not chordal. As a critical step in the algorithm for sparse triangular decomposition, we need to compute the chordal completions of these 22 associated graphs to find the perfect elimination orderings to use. According to Definition 2, we want to add as few edges in the chordal completion as possible to control the variable sparsity of the associated graphs.

For a graph $G = (V, E)$, if a set F of edges satisfies that $G' = (V, E \cup F)$ is chordal, then F is called a *triangulation* of G. Let F be a triangulation of G. If any proper subset of F is not a triangulation, then the triangulation F is said to be minimal. Existing algorithms for chordal completion mainly aim at minimal triangulation [12], and there are four typical algorithms for chordal completion,

Table 2. Variable sparsity of polynomial sets of ≥ 30 variables

Model	n	Chor	s_v	Model	n	Chor	s_v	Model	n	Chor	s_v	Model	n	Chor	s_v
014	86	0	0.746	333	54	0	0.395	478	33	0	0.301	584	35	1	0.005
105	39	0	0.336	334	74	0	0.376	491	57	0	0.244	594	33	0	0.693
205	194	0	0.114	335	34	0	0.303	492	52	0	0.293	635	80	0	0.219
220	58	0	0.132	362	34	0	0.323	501	35	0	0.398	636	78	0	0.231
270	33	0	0.138	407	47	0	0.157	504	75	0	0.110	667	103	0	0.229
332	78	0	0.355	416	32	1	0.117	559	90	0	0.207	705	43	0	0.143

namely the Lex-M [27], MCS-M [3] (these two algorithms are based on the Lex-BFS and MCS algorithms for recognizing chordality respectively), SMS (saturate minimal separators) [25], and CMT (clique minimal triangulation) [19] ones, with different strategies. To our knowledge, the Lex-M and MCS-M algorithms have the same computational complexity $O(mn)$, where n is the number of vertices and m is the number of edges (the last two algorithms above have complexity in complicated form so we do not detail here). Since m is bounded by $\binom{2}{n}$, the number of edges of a complete graph of n vertices, the complexity above is at most polynomial in the number of vertices n, which in our case is also equal to the number of variables in the polynomial set. Comparing this complexity with the (at least) exponential one for triangular decomposition, we can safely conclude that applying chordal completion with these two algorithms does not greatly affect the total computation time with sparse triangular decomposition.

We apply chordal completion to the 22 non-chordal associated graphs with existing implementations of all the four algorithms above. The variable sparsity of the polynomial sets after chordal completion is recorded in Table 3. In this table, the columns "s'_{v1}", "s'_{v2}", "s'_{v3}", and "s'_{v4}" record the variable sparsity after chordal completion by the Lex-M, MCS-M, SMS, and CMT algorithms respectively ("—" means that computation does not finish within 1 h), with the minimal sparsity among the 4 columns boxed, and the column "Inc" records the increase ratio of the minimal variable sparsity versus the original one.

As can be found from Table 3, among the four tested algorithms for chordal completion, the SMS algorithm returns chordal completions of the smallest sizes for 18 out of 22 (around 82%) associated graphs, while the Lex-M and MCS-M algorithms return marginally larger chordal completions. As regards the computation time (not recorded in the table), the Lex-M and MCS-M algorithms finish computation for all graphs within 10 and 5 s respectively, while the computation time of the SMS and CMT algorithms can go up to hundreds of seconds. Take both the output size and efficiency into consideration, based on our experiments we recommend the MCS-M algorithm for chordal completion of associated graphs of polynomial sets in biological dynamical systems.

After chordal completion, 9 out of 22 (round 41%) systems have increase ratios greater than 20% in the variable sparsity. The greatest increase occurs in Model 205, from 0.114 to 0.238 (108.61% increase ratio). We can conclude that chordal completion has a remarkable influence on the variable sparsity. However, even after chordal completion, the majority of systems are still relatively sparse

Table 3. Variable sparsity of polynomial sets of ≥ 30 variables after chordal completion

Model	n	s_v	s'_{v1}	s'_{v2}	s'_{v3}	s'_{v4}	Inc	Model	n	s_v	s'_{v1}	s'_{v2}	s'_{v3}	s'_{v4}	Inc
014	86	0.746	0.879	0.879	0.868	0.972	16.37%	478	33	0.301	0.320	0.320	0.320	0.608	6.29%
105	39	0.336	0.363	0.363	0.355	0.394	5.62%	491	57	0.244	0.385	0.347	0.326	0.873	33.59%
205	194	0.114	0.270	0.270	0.238	—	108.61%	492	52	0.293	0.465	0.454	0.437	0.898	48.84%
220	58	0.132	0.180	0.167	0.166	0.624	25.11%	501	35	0.398	0.403	0.403	0.403	0.627	1.27%
270	33	0.138	0.159	0.159	0.159	0.489	15.07%	504	75	0.110	0.138	0.138	0.138	0.712	25.25%
332	78	0.355	0.568	0.568	0.509	0.654	43.47%	559	90	0.207	0.209	0.209	0.209	0.255	0.84%
333	54	0.395	0.562	0.562	0.512	0.653	29.56%	594	33	0.693	0.703	0.703	0.703	0.735	1.37%
334	74	0.376	0.626	0.626	0.556	0.719	47.98%	635	80	0.219	0.224	0.221	0.222	0.641	0.72%
335	34	0.303	0.332	0.328	0.328	0.535	8.24%	636	78	0.231	0.236	0.232	0.233	0.688	0.62%
362	34	0.323	0.335	0.335	0.337	0.508	3.87%	667	103	0.229	0.357	0.357	0.367	0.867	56.03%
407	47	0.157	0.240	0.191	0.186	0.673	18.24%	705	43	0.143	0.159	0.159	0.156	0.175	9.30%

with respect to the variables (say, with variable sparsity less than 0.4) to furnish enough ones for our analysis on the influence of chordality on sparse and ordinary algorithms for triangular decomposition, to be reported in Sect. 4 below.

Take the following biological dynamical system labeled as BIOMD0000000270 in BioModels

$$
\begin{cases}
\dfrac{dx_1}{dt} = -\dfrac{122149}{10000000}x_1 x_{33} + \dfrac{5756}{15625}x_{10}x_{13}, \\[2mm]
\dfrac{dx_2}{dt} = -\dfrac{157857}{50000}x_{10}x_2 + \dfrac{23999}{20000}x_{11}x_{13}, \\[2mm]
\quad\vdots \\[2mm]
\dfrac{dx_{32}}{dt} = \dfrac{51051}{125000}x_{31} - \dfrac{51051}{125000}x_{32}, \\[2mm]
\dfrac{dx_{33}}{dt} = 0,
\end{cases}
\tag{2}
$$

for example. The associated graph of the polynomial set in the right hands above is shown in Fig. 1 (left), and it has 33 vertices and 73 edges, with the variable sparsity 0.138. With the MCS-M algorithm, the chordal completion (see Fig. 1 (right)) has 84 edges and variable sparsity 0.159. There is a 15.07% increase in the variable sparsity, but the graph after chordal completion is still sparse with respect to the variables.

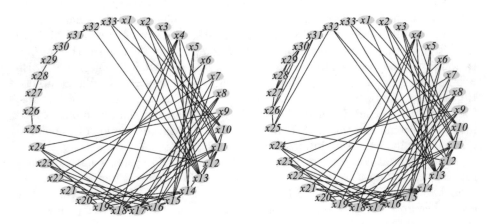

Fig. 1. Associated graph (left) and chordal completion (right) of polynomial set in (2)

4 Sparse Triangular Decomposition for Computing Equilibria

In this section, for those large polynomial sets whose variable sparsity is below 0.4 after chordal completion, we apply both the sparse and ordinary algorithms for regular decomposition and compare their computation time to reveal whether chordality can help to speed up triangular decomposition of sparse polynomial sets arising from biological dynamical systems. All the experiments were carried out on a Macbook Pro laptop with a 2.7 GHz dual core i5 CPU and 8GB 1867 MHz DDR3 memory under the operating system MacOS Sierra 10.15.6. We used the REGSER function for computing regular decomposition in top-down style for both sparse and ordinary algorithms in the EPSILON package for triangular decomposition in the Computer Algebra System MAPLE (Version 2018).

We tested all the 141 biological dynamical systems with ordinary regular decomposition to find that for 107 of them the computation finishes within 10 s. For such easy systems using sparse regular decomposition does not make much sense, for the time saving would not be remarkable. For the remaining systems, 17 of them are of variable sparsity lower than 0.6 after chordal completion, so we choose them for further comparisons.

For these 17 biological dynamical systems, for the sparse algorithm we first construct the chordal completions in case the associated graphs are not chordal, and then extract three perfect elimination orderings as variable orderings for computation with sparse regular decomposition, while for the ordinary regular decomposition three variable orderings are randomly generated and then verified not to be perfect elimination orderings. The experimental results for these 17 systems are summarized in Table 4. The columns "t_{p1}", "t_{p2}", and "t_{p3}" record the CPU time in seconds for sparse regular decomposition with three chosen perfect elimination orderings ("—" means that the computation does not finish within 2 h), the columns "t_{r1}", "t_{r2}", and "t_{r3}" record that for ordinary regular decomposition with three random variable orderings, the columns "$\overline{t_p}$" and "$\overline{t_r}$" are the average time for three perfect elimination orderings and random variable orderings respectively, and the average speedup ratios of sparse versus ordinary regular decomposition is recorded in column "Speedup".

Take the system labeled as BIOMD0000000504 in BioModels for example. It is a large system involving 75 variables and also sparse with respect to them (its variable sparsity before and after chordal completion is 0.110 and 0.138 respectively). One perfect elimination ordering ("t_{p1}" in Table 4) obtained from the chordal completion of its associated graph is shown below (as a sequence of subscripts of the variables).

$$11, 25, 28, 29, 71, 60, 10, 72, 63, 26, 62, 52, 64, 61, 27, 31, 30, 65, 3, 73, 6, 2, 66, 20, 67,$$
$$15, 8, 68, 46, 45, 43, 48, 36, 51, 34, 49, 54, 53, 50, 47, 44, 35, 69, 70, 13, 4, 5, 32, 21, 33, \quad (3)$$
$$9, 14, 22, 23, 24, 16, 18, 1, 17, 12, 19, 57, 7, 55, 41, 38, 42, 58, 56, 59, 75, 39, 40, 74, 37.$$

With respect to this variable ordering, the REGSER function returns 90 triangular sets, with one simple triangular set as

$$[x_{11}, 400000x_{25}^3 + 505080x_{25}^2 + 101x_{25}, -20000x_{29} + 5000x_{25}^2 + x_{25}, 100x_{71} + 80x_{25}x_{29} + x_{29}, x_{60},$$

$$-2x_{72} + x_{25}, -8x_{26} + 75x_{25}, -2x_{62} + 9375x_{25}x_{63}, x_{52}, x_{61}, -2x_{27} + 15x_{25}, -7x_{31} + 10x_{25} + 140,$$

$$x_{65}, 6x_3 - 35x_{31} + 700, 6x_3x_{73} + 350x_{73} - 35x_{30}, x_6, x_2 - 5x_3, x_{66}, x_{20}, x_{67}, x_{15}, x_8, x_{68}, x_{46}, x_{45},$$

$$x_{43}, x_{48}, x_{36}, x_{51}, x_{34}, x_{49}, -7x_{54} + 24x_3 + 2800, -7x_{53} + 12x_3x_{73} + 700x_{73}, x_{50}, x_{47}, x_{44}, x_{35},$$

$$x_{69}, x_{70}, x_{13}, x_4 - 100, x_5 - 5, x_{21}, x_{33}, x_9, x_{14}, x_{16}, x_{18}, x_1, x_{17}, x_{57}, x_{55}, x_{42}, x_{56}].$$

We denote this triangular set by $\mathcal{T}$, and it consists of 54 polynomials whose leading variables (the first variable in the each corresponding expression above) strictly increase following the variable order specified in (3). It is easy to see that the polynomials in $\mathcal{T}$ are very simple in their form: they involve no more than 3 variables, and their degrees with respect to the leading variables are low. In fact, many polynomials are linear in their leading variables and even as simple as one variable only.

Comparing the numbers of variables (75) and polynomials (54) in $\mathcal{T}$, we know that 21 of the variables do not have corresponding polynomials in $\mathcal{T}$ with them as leading variables, and thus in the process of successively solving the polynomial equation set $\mathcal{T} = 0$ defined by $\mathcal{T}$, these 21 variables are considered as *parameters* who can take arbitrary values. For example, with the polynomial equation $400000x_{25}^3 + 505080x_{25}^2 + 101x_{25} = 0$ in $\mathcal{T} = 0$, we are able to compute the values of x_{25}. Since there is no polynomial in $\mathcal{T}$ with leading variable x_{63}, we know that x_{63} is a parameter. With the values of x_{25} solved above and of x_{63} specified, we can easily compute the values of x_{62} with $-2x_{62} + 9375x_{25}x_{63} = 0$ in $\mathcal{T} = 0$. In this way, one is able to find all the solutions of $\mathcal{T} = 0$.

By solving the remaining triangular sets in the same way as above and combining all the solution sets of the 90 triangular sets, we are able to explicitly and symbolically compute all the solutions of the polynomial system, namely all the equilibria of the biological dynamical system, from which further stability analysis can be conducted.

From Table 4 one can find that for all the systems whose computation finishes within 2 h, sparse regular decomposition is superior in computational performances than the ordinary one, with the average speed-up ratios between 1.53 to 3.39. This means that variable sparsity in the biological dynamical systems are indeed utilized to improve the computational performances in sparse regular decomposition. In particular, for the system labeled "501", the sparse regular decomposition can return the results very quickly but the ordinary one cannot within 2 h for any of the three random variable orderings.

By comparing the three chosen perfect elimination orderings for sparse regular decomposition, one can see that the computation time also varies remarkably. This indicates that from the computational view of point, different perfect elimination orderings for the same chordal graph can have remarkable influence on the computation time of sparse triangular decomposition, and it remains as further study which perfect elimination ordering is better and why.

Table 4. Computation time for sparse and ordinary regular decomposition

Model	n	s'_v	t_{p1}	t_{p2}	t_{p3}	t_{r1}	t_{r2}	t_{r3}	$\overline{t_p}$	$\overline{t_r}$	Speedup
160	25	0.200	2.793	7.420	119.269	10.462	24.936	288.332	43.161	107.910	2.50
205	194	0.238	—	—	—	—	—	—			
220	58	0.166	37.263	125.788	435.904	240.095	477.932	602.467	199.652	440.165	2.20
332	78	0.509	189.764	541.506	986.204	673.333	1201.804	1540.605	572.491	1138.581	1.99
333	54	0.512	26.389	36.978	41.446	46.035	99.974	153.971	34.938	99.993	2.86
334	74	0.556	238.222	430.431	1073.222	687.416	757.550	1219.345	580.625	888.104	1.53
335	34	0.328	7.139	10.467	18.798	19.210	30.354	33.955	12.135	27.840	2.29
362	34	0.335	11.269	26.596	64.181	69.546	70.239	71.055	34.015	70.280	2.07
407	47	0.186	—	—	—	—	—	—			
478	33	0.320	2.047	3.737	5.49	8.947	9.934	19.355	3.758	12.745	3.39
501	35	0.403	4.561	7.483	12.954	—	—	—	8.333		
504	75	0.138	61.849	90.391	224.201	157.079	323.271	440.134	125.480	306.828	2.45
599	30	0.538	—	—	—	—	—	—			
635	80	0.221	—	—	—	—	—	—			
636	78	0.232	—	—	—	—	—	—			
667	103	0.357	—	—	—	—	—	—			
705	43	0.156	—	—	—	—	—	—			

5   Concluding Remarks

In this paper we apply sparse triangular decomposition to detect the equilibria of biological dynamical systems to exploit their inherent variable sparsity. As an essential step in sparse triangular decomposition, chordal completion is shown experimentally to incur a remarkable increase of the variable sparsity of such systems, and the MCS-M algorithm for chordal completion is recommended due to its overall practical performances. For large biological dynamical systems which are sparse with respect to their variables, the experimental results verify the efficiency gains of sparse triangular decomposition versus the ordinary one, with the speedup ratios between 1.5−3.4.

References

1. Allen, L.J.: Some discrete-time SI, SIR, and SIS epidemic models. Math. Biosci. **124**(1), 83–105 (1994)
2. Aubry, P., Lazard, D., Moreno Maza, M.: On the theories of triangular sets. J. Symbolic Comput. **28**(1–2), 105–124 (1999)
3. Berry, A., Blair, J.R.S., Heggernes, P., Peyton, B.W.: Maximum cardinality search for computing minimal triangulations of graphs. Algorithmica **39**(4), 287–298 (2004)
4. Bodlaender, H., Koster, A.: Combinatorial optimization on graphs of bounded treewidth. Comput. J. **51**(3), 255–269 (2018)

5. Boulier, F., Lefranc, M., Lemaire, F., Morant, P.-E.: Applying a rigorous quasi-steady state approximation method for proving the absence of oscillations in models of genetic circuits. In: Horimoto, K., Regensburger, G., Rosenkranz, M., Yoshida, H. (eds.) AB 2008. LNCS, vol. 5147, pp. 56–64. Springer, Heidelberg (2008). https://doi.org/10.1007/978-3-540-85101-1_5

6. Chen, C.: Chordality preserving incremental triangular decomposition and its implementation. In: Bigatti, A.M., Carette, J., Davenport, J.H., Joswig, M., de Wolff, T. (eds.) ICMS 2020. LNCS, vol. 12097, pp. 27–36. Springer, Cham (2020). https://doi.org/10.1007/978-3-030-52200-1_3

7. El Kahoui, M., Weber, A.: Deciding Hopf bifurcations by quantifier elimination in a software-component architecture. J. Symbolic Comput. **30**(2), 161–179 (2000)

8. Ferrell, J.E., Tsai, T.Y.C., Yang, Q.: Modeling the cell cycle: Why do certain circuits oscillate? Cell **144**(6), 874–885 (2011)

9. Galor, O.: Discrete Dynamical Systems. Springer, Heidelberg (2007). https://doi.org/10.1007/3-540-36776-4

10. Gatermann, K., Huber, B.: A family of sparse polynomial systems arising in chemical reaction systems. J. Symbolic Comput. **33**(3), 275–305 (2002)

11. Grigoriev, D., Iosif, A., Rahkooy, H., Sturm, T., Weber, A.: Efficiently and effectively recognizing toricity of steady state varieties. Preprint at arXiv:1910.04100 (2019)

12. Heggernes, P.: Minimal triangulations of graphs: A survey. Discret. Math. **306**(3), 297–317 (2006)

13. Hong, H., Liska, R., Steinberg, S.L.: Testing stability by quantifier elimination. J. Symbolic Comput. **24**(2), 161–187 (1997)

14. Laubenbacher, R., Sturmfels, B.: Computer algebra in systems biology. Amer. Math. Monthly **116**(10), 882–891 (2009)

15. Laubenbacher, R., Stigler, B.: A computational algebra approach to the reverse engineering of gene regulatory networks. J. Theor. Biol. **229**(4), 523–537 (2004)

16. Lauritzen, S., Spiegelhalter, D.: Local computations with probabilities on graphical structures and their application to expert systems. J. R. Stat. Soc: Series B. (Methodol.) **50**(2), 157–194 (1988)

17. Li, C., Donizelli, M., Rodriguez, N., Dharuri, H., et al.: BioModels database: An enhanced, curated and annotated resource for published quantitative kinetic models. BMC Syst. Biol. **4**(1), 92 (2010)

18. Li, X., Mou, C., Niu, W., Wang, D.: Stability analysis for discrete biological models using algebraic methods. Math. Comput. Sci. **5**(3), 247–262 (2011)

19. Mezzini, M., Moscarini, M.: Simple algorithms for minimal triangulation of a graph and backward selection of a decomposable Markov network. Theor. Comput. Sci. **411**(7–9), 958–966 (2010)

20. Mou, C.: Symbolic detection of steady states of autonomous differential biological systems by transformation into block triangular form. In: Jansson, J., Martín-Vide, C., Vega-Rodríguez, M.A. (eds.) AlCoB 2018. LNCS, vol. 10849, pp. 115–127. Springer, Cham (2018). https://doi.org/10.1007/978-3-319-91938-6_10

21. Mou, C., Bai, Y.: On the chordality of polynomial sets in triangular decomposition in top-down style. In: Proceedings of ISSAC 2018, pp. 287–294. ACM Press (2018)

22. Mou, C., Bai, Y., Lai, J.: Chordal graphs in triangular decomposition in top-down style. J. Symbolic Comput. **102**, 108–131 (2021)

23. Niu, W., Wang, D.: Algebraic approaches to stability analysis of biological systems. Math. Comput. Sci. **1**(3), 507–539 (2008)

24. Niu, W., Wang, D.: Algebraic analysis of bifurcation and limit cycles for biological systems. In: Horimoto, K., Regensburger, G., Rosenkranz, M., Yoshida, H. (eds.) AB 2008. LNCS, vol. 5147, pp. 156–171. Springer, Heidelberg (2008). https://doi. org/10.1007/978-3-540-85101-1_12
25. Parra, A., Scheffler, P.: Characterizations and algorithmic applications of chordal graph embeddings. Discret. Appl. Math. **79**(1–3), 171–188 (1997)
26. Rose, D.: Triangulated graphs and the elimination process. J. Math. Anal. Appl. **32**(3), 597–609 (1970)
27. Rose, D., Tarjan, E., Lueker, G.: Algorithmic aspects of vertex elimination on graphs. SIAM J. Comput. **5**(2), 266–283 (1976)
28. Sturm, T., Weber, A., Abdel-Rahman, E., El Kahoui, M.: Investigating algebraic and logical algorithms to solve Hopf bifurcation problems in algebraic biology. Math. Comput. Sci. **2**(3), 493–515 (2009)
29. Wang, D., Xia, B.: Stability analysis of biological systems with real solution classification. In: Proceedings of ISSAC 2005, pp. 354–361. ACM Press (2005)
30. Wang, D.: Computing triangular systems and regular systems. J. Symbolic Comput. **30**(2), 221–236 (2000)
31. Wang, D.: Elimination Methods. Springer, Vienna (2001). https://doi.org/10. 1007/978-3-7091-6202-6

A Recovery Algorithm and Pooling Designs for One-Stage Noisy Group Testing Under the Probabilistic Framework

Yining Liu, Sachin Kadyan, and Itsik Pe'er[✉]

Department of Computer Science, Columbia University, New York, USA
{yl4536,sk4835,ip2169}@columbia.edu

Abstract. Group testing saves time and resources by testing each pre-assigned group instead of each individual, and one-stage group testing emerged as essential for cost-effectively controlling the current COVID-19 pandemic. Yet, the practical challenge of adjusting pooling designs based on infection rate has not been systematically addressed. In particular, there are both theoretical interests and practical motivation to analyze one-stage group testing at finite, practical problem sizes, rather than asymptotic ones, under noisy, rather than perfect tests, and when the number of positives is randomly distributed, rather than fixed.

Here, we study noisy group testing under the probabilistic framework by modelling the infection vector as a random vector with Bernoulli entries. Our main contributions include a practical one-stage group testing protocol guided by maximizing pool entropy and a maximum-likelihood recovery algorithm under the probabilistic framework. Our findings highlight the implications of introducing randomness to the infection vectors – we find that the combinatorial structure of the pooling designs plays a less important role than the parameters such as pool size and redundancy.

Keywords: Non-adaptive group testing · COVID-19 · Experimental designs

1 Introduction

Group testing is a procedure to find positives in a cohort by applying tests for the presence of any positives to cohort subsets (groups), instead of testing each individual separately. When the fraction of positives among the samples is low, implementing group testing saves time and resources. Group testing has applications in genetics [21], drug screening [13] communications [25] and epidemiology [23].

Columbia University unrestricted funds.

C. Martín-Vide et al. (Eds.): AlCoB 2021, LNBI 12715, pp. 42–53, 2021.
https://doi.org/10.1007/978-3-030-74432-8_4

One-stage group testing [6] addresses the scenario of groups being set in advance, independently of test results - a common practical requirement, e.g. due to testing speed constraints. Existing algorithms on infection vector recovery for noisy group testing under the combinatorial prior (the number of positives among samples is fixed) include LP relaxation [16], belief propagation [20], Markov Chain Monte Carlo (MCMC) [9], Noisy Combinatorial Orthogonal Matching Pursuit (NCOMP) [4], separate decoding [18] and Definite positives (DD) [19]. However, group testing under a fully probabilistic framework has not been extensively studied. In particular, to the best of our knowledge, there has been a lack of work on non-asymptotic results on noisy group testing in the realistic scenario where number of positives in not known in advance, as in [2,3], but rather is a random variable.

As part of the global efforts to control the COVID-19 pandemic, there has been an emerging body of work on implementing one-stage group testing for COVID-19 [10,11,22–24]. However, testing scenarios vary significantly due to fluctuating infection rates across time and geography often caused by emerging variants, as well as different testing objectives, such as screening of health care workers versus large scale monitoring of the community. Group testing protocols should thus be adjusted according to the infection rate among the tested samples, measurement error rates, and the recovery error tolerance level [5].

As a result, systematically studying one-stage noisy group testing with a random number of positives is of both theoretical interests and practical importance. In this paper, we study noisy group testing under the probabilistic framework in order to address the practical challenges posted by group testing implementation for COVID-19.

The rest of the paper is organized as follows. We begin with an introduction of the noisy group testing under the probabilistic framework in Sect. 2, including a group testing protocol and a novel recovery algorithm under the probabilistic framework. The performance of the recovery algorithm and the pooling designs is in Sect. 3. Finally, we conclude with a discussion on future work in Sect. 4.

2 Methods

2.1 Noisy Group Testing Under the Probabilistic Framework

We assume a tested cohort of n individuals, with some infection rate f among the population the cohort is sampled for. Note, that the actual number of true positives is not known in advance, as combinatorial priors unrealistically assume [2,3]. Also, this is the ground truth rate of infected individuals, as opposed to the observed positivity rate. Specifically:

Definition 1. *An infection vector X is a random vector with n i.i.d. Bernoulli(f) entries.*

We aim to design a protocol to test n samples with $t < n$ tests. We arrange the pool assignments into a pooling matrix $M \in \{0,1\}^{t \times n}$ such that $M_{ij} = 1$ if and only if individual j is included in pool i. Notice that each row i sum of M corresponds to a pool size s_i, and each column j sum corresponds to the number of pools that a sample participates, which we define as redundancy r_j.

Following [1] and [8], we focus on the following four classes of pooling designs that are of practical interests (see Fig. 1):

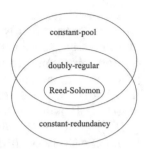

Fig. 1. Four classes of pooling designs.

1. (constant-pool design) each test combines a fixed number $s_i = s$ of samples, where s samples are chosen uniformly at random among the n samples;
2. (constant-redundancy design) each individual participates in a fixed number $r_j = r$ of tests, where r tests are chosen uniformly at random among the t tests;
3. (doubly-regular design) draw a design uniformly at random from the designs that are both constant-pool and constant-redundancy;
4. (Reed-Solomon design) the pool assignment for each individual is obtained as a concatenation of Reed-Solomon error correcting code; see [8] for more details on the explicit construction.

Under the noiseless setting, a test result is negative if and only if all samples in the pool are negative. In practice, test results might suffer from measurement errors. Suppose the test we use for each pool has a false negative rate of f_0 and a false positive rate of f_1, i.e. consider the asymmetric noisy channel (where f_0 and f_1 can be distinct) that combines the additive model and the dilution model in [2]. See Fig. 2 for an example of a doubly-regular pooling matrix under the probabilistic framework.

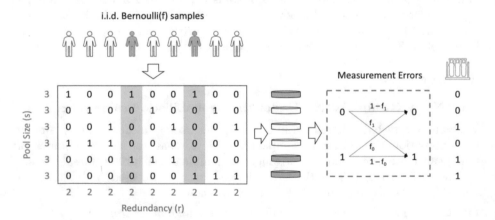

Fig. 2. Noisy group testing under the probabilistic framework. The pooling matrix is a doubly-regular design with $s = 3$ and $r = 2$. A pool is negative if and only if there is no positive sample participates in the pool. Each pool is then passed to an asymmetric noisy channel to model measurement errors.

2.2 A Group Testing Protocol

Let Y_i be the indicator that the test result for the pool i is negative; then

$$Y_i \sim \text{Bernoulli}(f_0(1 - (1 - f)^{s_i}) + (1 - f_1)(1 - f)^{s_i}).$$

Extending the idea in [22] to the noisy setting, a pool design should maximize the entropy of Y_i; hence the optimal pool size is

$$\arg \max_{s_i} H(Y_i; s_i) = \frac{\ln(\frac{0.5 - f_0}{1 - f_1 - f_0})}{\ln(1 - f)}. \tag{1}$$

Infection rates therefore dramatically affect the theoretical optimal pool size, while measurement errors slightly offsets such optima (see Fig. 3).

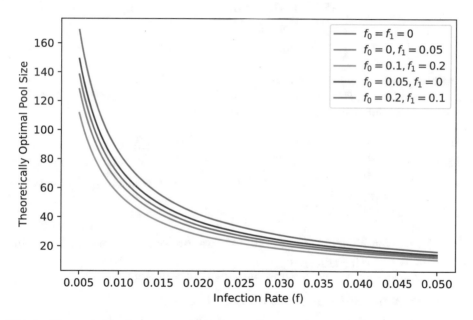

Fig. 3. Theoretically optimal pool size as a function of infection rate; false positives (resp. negatives) slightly reduce (increase) optimal pool size.

We will show in Sect. 3 that given a fixed set of parameters r, s, t, the four classes of designs (and that of each member of a given class) perform comparably under this probabilistic framework. Hence, practitioners can safely choose a design class with practical advantages and then optimize the parameters r, s, t.

We thus propose the following protocol for implementing one-stage group testing under a given infection rate:

1. From the potential pool sizes (subject to practical constraints), choose s to be closest to the theoretical optimal pool size, per Eq. 1.
2. Simulate the performance curve using the recovery algorithm in Sect. 2.3.
3. Choose the number of pools t based on the error tolerance level.
4. Perform one-stage group testing with the same recovery algorithm in step 2.

2.3   Recovery Algorithm

We now introduce an algorithm to recover the infection vector for noisy group testing under the probabilistic framework.

Assume we are given a pooling design M with t pools; the pool results is

$$Y := \text{sign}(MX) - E^{(0)} + E^{(1)}$$

where $E^{(0)}, E^{(1)}$ are false-negatives vector and false-positives vector respectively, i.e. for $i = 1, \ldots, t$,

$$
\begin{cases}
E_i^{(0)} \sim \text{Bernoulli}(f_0) & \text{if } (MX)_i > 0, \\
E_i^{(0)} \sim \text{Bernoulli}(0) & \text{if } (MX)_i = 0, \\
E_i^{(1)} \sim \text{Bernoulli}(f_1) & \text{if } (MX)_i = 0, \\
E_i^{(1)} \sim \text{Bernoulli}(0) & \text{if } (MX)_i > 0.
\end{cases}
$$

A recovery algorithm is given a pool result $y \in \{0,1\}^t$ and outputs $(\hat{x}, \hat{e}^{(0)}, \hat{e}^{(1)})$, an estimate for $(X, E^{(0)}, E^{(1)})$. Our recovery algorithm (**Algorithm 1**) is given by solving the following integer linear program:

$$
\text{minimize} \quad -||\hat{x}|| \ln(\tfrac{f}{1-f}) - ||\hat{e}^{(0)}|| \ln(\tfrac{f_0}{1-f_0}) - ||\hat{e}^{(1)}|| \ln(\tfrac{f_1}{1-f_1})
$$

$$
\begin{aligned}
\text{subject to } 0.5\hat{e}_i^{(0)} &\leq (M\hat{x})_i \leq t\hat{e}_i^{(0)} + 0.5 & i \in \{1,\ldots,t\} \text{ s.t. } y_i = 0 \\
-0.5\hat{e}_i^{(1)} + 0.5 &\leq (M\hat{x})_i \leq t(1 - \hat{e}_i^{(1)}) + 0.5 & i \in \{1,\ldots,t\} \text{ s.t. } y_i = 1 \\
\hat{x} &\in \{0,1\}^n \\
\hat{e}^{(0)}, \hat{e}^{(1)} &\in \{0,1\}^t
\end{aligned}
$$

Here, for a given binary vector $x \in \{0,1\}^n$, we use $||x||$ to denote the Hamming weight of x, i.e. $||x|| = \sum_{i=1}^{n} x_i$.

We now analyze the correctness and optimality of Algorithm 1.

Theorem 2. *Algorithm 1 returns a realizable output.*

Proof. In order for an output $(\hat{x}, \hat{e}^{(0)}, \hat{e}^{(1)})$ to be realizable, each row of the equation $y = \text{sign}(M\hat{x}) - \hat{e}^{(0)} + \hat{e}^{(1)}$ must fall into one of the following cases:

$$
\begin{cases}
y_i = 0, (M\hat{x})_i = 0, \hat{e}_i^{(0)} = 0; \\
y_i = 0, (M\hat{x})_i > 0, \hat{e}_i^{(0)} = 1; \\
y_i = 1, (M\hat{x})_i > 0, \hat{e}_i^{(1)} = 0; \\
y_i = 1, (M\hat{x})_i = 0, \hat{e}_i^{(1)} = 1.
\end{cases}
$$

Notice that for $y_i = 0$, we have

$$
\hat{e}_i^{(0)} = 1 \iff 0.5 \leq (M\hat{x})_i \leq t + 0.5 \iff 1 \leq (M\hat{x})_i \leq t,
$$

$$
\hat{e}_i^{(0)} = 0 \iff 0 \leq (M\hat{x})_i \leq 0.5 \iff (M\hat{x})_i = 0.
$$

Similarly, for $y_i = 1$, we have

$$\hat{e}_i^{(1)} = 0 \iff 0.5 \le (M\hat{x})_i \le t + 0.5 \iff 1 \le (M\hat{x})_i \le t,$$
$$\hat{e}_i^{(1)} = 1 \iff 0 \le (M\hat{x})_i \le 0.5 \iff (M\hat{x})_i = 0.$$

Therefore, the constraints in the integer linear program guarantees the output is realizable.

Theorem 3. *Algorithm 1 returns a maximum likelihood estimate (MLE), i.e. an output $(\hat{x}, \hat{e}^{(0)}, \hat{e}^{(1)})$ maximizes $\Pr(X = x, E^{(0)} = e^{(0)}, E^{(1)} = e^{(1)})$.*

Proof. Let n_0 be the number of pools with no positive sample and n_1 be the number of pools containing at least one positive sample. The log-likelihood is

$$\ln \Pr(X = x, E^{(0)} = e^{(0)}, E^{(1)} = e^{(1)})$$
$$= \ln \Pr(X = x) + \ln \Pr(E^{(0)} = e^{(0)}, E^{(1)} = e^{(1)} | X = x)$$
$$= ||x|| \ln f + (n - ||x||) \ln(1 - f) + ||e^{(0)}|| \ln f_0 + (n_1 - ||e^{(0)}||) \ln(1 - f_0)$$
$$\quad + ||e^{(1)}|| \ln f_1 + (n_0 - ||e^{(1)}||) \ln(1 - f_1)$$
$$= ||x|| \ln(\frac{f}{1 - f}) + ||e^{(0)}|| \ln(\frac{f_0}{1 - f_0}) + ||e^{(1)}|| \ln(\frac{f_1}{1 - f_1}) + \text{const}.$$

Therefore, the objective function in the integer linear program maximizes the likelihood.

Open source code implementing all our methods is available at [15].

3   Results

For biological experiments, sample sizes $n = 96, 384, 1536$ are of particular interests because the experiments tend to be conducted on 96-well plates, 384-well plates, or 1536-well plates [8]. The savings in resources offered by pooled design compared to individual testing stops being worth the increased complexity of the experimental paradigm when infection rates are high. For example, under an infection rate $f = 0.1$, using a constant-pool design with optimal pool size $s = 7$, even a 2-fold saving, with $T = n/2 = 192$ pools to recover $n = 384$ samples, the average accuracy over 1000 trials is unacceptably low at 96.5%. As a result, we believe our group testing protocol should not be applied when the infection rate is over 0.1, and limit our analysis to infection rate up to 0.025 in this section.

3.1   Performance of the Recovery Algorithm

In order to test the performance of the recovery algorithm, we used the pooling design $M \in \{0, 1\}^{48 \times 384}$ (testing 384 individuals in 48 pools) in [23].

We simulated 1000 infection vectors $x \in \{0, 1\}^{384}$ under the population infection rate $f = 0.02$ and evaluated recovery under different rates of false positives and false negatives. We observe error rate to only slightly detract from the accuracy of noiseless measurements (see Fig. 4).

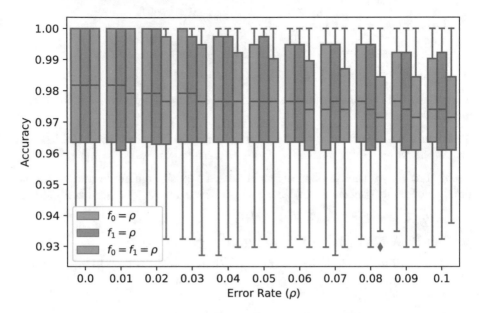

Fig. 4. Robustness against measurements errors.

3.2   Pooling Matrices

Optimal Pool Size. In order to test the theoretically optimal pool size given in Eq. 1, we simulated 1000 infection vectors under various infection rates and evaluated the performance of constant-pool matrix with $t = 48$ and $s \in \{8, 16, 24, \ldots, 80\}$. The performance of the pooling matrices is strongly correlated with the entropy of each pool, which supports the guidance of maximizing pool entropy in one-stage group testing designs (see Fig. 5).

In order to test the robustness of the optimal pool size s with respect to varying the number of pools t, we tested the performance of constant-pool designs with $s \in \{8, 16, 24, \ldots, 80\}$ based on 1000 simulated infection vectors under $f = 0.02$. The optimal pool size ($s = 32$) achieves nearly maximal recovery accuracy for all t (see Fig. 6).

Explicit Constructions. Combinatorial structure of pooling matrices has been studied under the combinatorial framework [7]. Recent development of explicit pooling matrices construction for noisy group testing is based on error correcting codes [1]. Pooling design based on Reed-Solomon code was introduced in [14] and shown to be optimal under a more restricted probabilistic setting (assuming a random set of positives of a fixed size) [12].

Authors in [23] use a design based on Reed-Solomon error correcting code; the matrix assigns each individual in 6 pools such that each pool has 48 individuals. We compare the performance of the Reed-Solomon design, a randomly drawn doubly-regular design (with $r = 6$ and $s = 48$), a randomly drawn

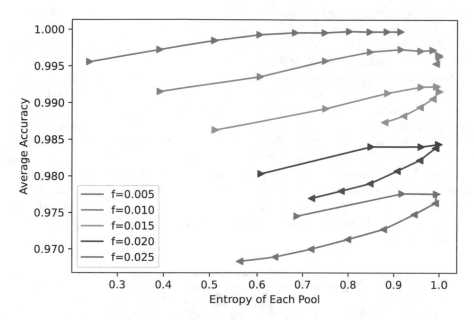

Fig. 5. Entropy of each pool against the performance of constant-pool designs with pool size $s \in \{8, 16, \ldots, 80\}$ and $t = 48$ pools; infection vectors are simulated with 384 Bernoulli(f) entries. The direction of the triangular markers indicates increasing s.

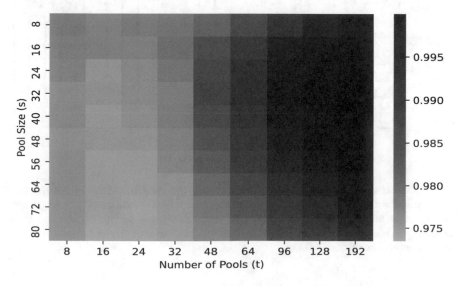

Fig. 6. Average recovery accuracy of designs with the pool size $s \in \{8, 16, 24, \ldots, 80\}$ and varying the number of pools under the infection rate $f = 0.02$; the pool size $s = 32$ achieves the maximum pool entropy.

constant-redundancy design (with $r = 6$ and varying pool sizes), and a randomly drawn constant-pool design (with $s = 48$ and varying redundancies).

We found that under the probabilistic framework, the performance of the three random designs is comparable to that of the Reed-Solomon design in terms of both the average and the standard deviation of the recovery accuracy (see Fig. 7). For each of the three random designs, we computed the distribution of the inner products of column vectors (which are derived from code words) of 1000 randomly-drawn matrices. The performance guarantees of Reed-Solomon design under the combinatorial setting follows from the minimum pairwise distance of the code words [1]. The authors in [17] studied explicit pooling matrices and showed the dependency of the error probability on both the minimum and the average distance under the assumption of a fixed number of positives. In the same spirit, we found that maximizing the number of pairs of column vectors with zero inner products could lead to performance guarantees under the probabilistic framework (see Fig. 8).

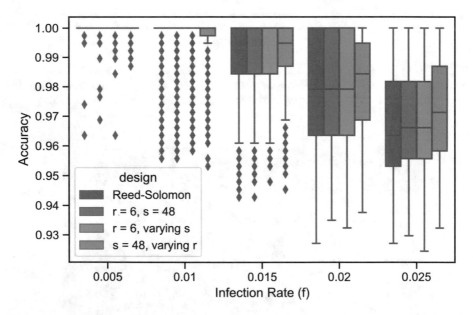

Fig. 7. Performance of the Reed-Solomon (red), the doubly-regular $r = 6, s = 48$ (blue), the constant redundancy $r = 6$ (green), and the constant pool $s = 48$ (orange) designs under simulated data with $n = 384$ individuals, $t = 48$ pools. The box shows the four quartiles, the whiskers show the rest of the distribution, and the points show the outliers. (Color figure online)

As a first step towards understanding the role of combinatorial structures of the pooling matrices under the probabilistic framework, we randomly drew 1000 constant-pool designs with $s = 48$ to test $n = 384$ individuals with $t = 48$ pools, and computed the average accuracy on two sets of 1000 simulated infection

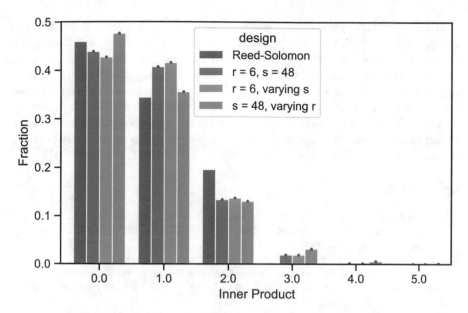

Fig. 8. Distributions of the inner products of column vector. Error bars are computed based on 1000 matrices.

vectors under the infection rate $f = 0.02$. We observe low correlation (≈ 0.0884; $p \approx 0.00516$) between the performance of designs on the two sets of vectors, which suggests that under the probabilistic framework, the combinatorial structure of the pooling matrices plays a less significant role than the parameters s, r, t.

4   Discussion

We presented a framework that addresses several practical issues, including randomness of the infection vector and noisy measurements. Our findings suggest that in practice, the combinatorial structure of pooling design matrices plays a lesser role compared to parameters such as redundancy, pool sizes, and the number of pools. Furthermore, we provide a protocol and an implementation for practitioners to choose the parameters in order to design pooling matrices.

From a theoretical perspective, there are several promising future directions for this work. The ILP problem presented in Sect. 2.3 can be viewed as a generalization of the smallest satisfying set (SSS) problem [1], which is NP-hard. One direction of future work is to design polynomial-time approximation recovery algorithm under the probabilistic framework.

In addition, non-asymptotic theoretical understanding of one-stage group testing under the probabilistic framework is still limited. An interesting future research direction is to explore theoretical justifications on the effect of the number of pairs of column vectors with zero inner product on the performance, as observed in Sect. 3.

From a practical perspective, while no significant difference in performance among the four classes of designs is observed in the simulations performed in this study, future research should be conducted on further investigating the significance of the combinatorial structure of the pooling matrices under other scenarios such that parameter values with a higher range of recovery accuracy and sample sizes other than $n = 384$ that are germane to biological experiments.

As the COVID-19 pandemic continue to ravage large populations of resource limited nations, cost effective testing with simple designs is an absolute necessity for sustaining daily life in a safe manner.

References

1. Aldridge, M., Johnson, O., Scarlett, J.: Group testing: an information theory perspective. Found. Trends Commun. Inf. Theory **15**(3–4), 196–392 (2019). https://doi.org/10.1561/0100000099
2. Atia, G., Saligrama, V.: Noisy group testing: an information theoretic perspective. In: 2009 47th Annual Allerton Conference on Communication, Control, and Computing (Allerton), pp. 355–362. IEEE (2009). https://doi.org/10.1109/ALLERTON.2009.5394787
3. Atia, G.K., Saligrama, V.: Boolean compressed sensing and noisy group testing. IEEE Trans. Inf. Theory **58**(3), 1880–1901 (2012). https://doi.org/10.1109/TIT.2011.2178156
4. Chan, C.L., Che, P.H., Jaggi, S., Saligrama, V.: Non-adaptive probabilistic group testing with noisy measurements: near-optimal bounds with efficient algorithms. In: 2011 49th Annual Allerton Conference on Communication, Control, and Computing (Allerton), pp. 1832–1839. IEEE (2011). https://doi.org/10.1109/Allerton.2011.6120391
5. Deka, S., Kalita, D.: Effectiveness of sample pooling strategies for SARS-CoV-2 mass screening by RT-PCR: a scoping review. J. Lab. Phys. **12**(03), 212–218 (2020). https://doi.org/10.1055/s-0040-1721159
6. Du, D.: Pooling Designs and Nonadaptive Group Testing: Important Tools for DNA Sequencing. World Scientific Publishing Company, New Jersey (2006)
7. Du, D., Hwang, F.: Combinatorial Group Testing and Its Applications. World Scientific Publishing Company, Singapore (1993)
8. Erlich, Y., et al.: Biological screens from linear codes: theory and tools (2015). https://www.biorxiv.org/content/10.1101/035352v1.article-info. Accessed 14 Jan 2021
9. Furon, T., Guyader, A., Cérou, F.: Decoding fingerprints using the Markov chain Monte Carlo method. In: 2012 IEEE International Workshop on Information Forensics and Security (WIFS), pp. 187–192 (2012). https://doi.org/10.1109/WIFS.2012.6412647
10. Ghosh, S., et al.: A compressed sensing approach to group-testing for COVID-19 detection. https://arxiv.org/abs/2005.07895 (2020). Accessed 14 Jan 2021
11. Ghosh, S., et al.: Tapestry: a single-round smart pooling technique for COVID-19 testing (2020). https://www.medrxiv.org/content/10.1101/2020.04.23.20077727v2. Accessed 14 Jan 2021
12. Inan, H.A., Kairouz, P., Wootters, M., Ozgur, A.: On the optimality of the Kautz-singleton construction in probabilistic group testing. In: 56th Annual Allerton Conference on Communication, Control, and Computing (Allerton), pp. 188–195. IEEE (2018). https://doi.org/10.1109/ALLERTON.2018.8635972

13. Kainkaryam, R.M., Woolf, P.J.: Pooling in high-throughput drug screening. Curr. Opin. Drug Discov. Devel. **12**(3), 339–350 (2009)
14. Kautz, W.H., Singleton, R.C.: Nonrandom binary superimposed codes. IEEE Trans. Inf. Theory **10**(4), 363–377 (1964). https://doi.org/10.1109/TIT.1964.1053689
15. Liu, Y., Kadyan, S., Pe'er, I.: Group testing under the probabilistic framework code (2021). https://github.com/imyiningliu/group-testing-probabilistic-framework. Accesed 28 Feb 2021
16. Malioutov, D., Malyutov, M.: Boolean compressed sensing: LP relaxation for group testing. In: 2012 IEEE International Conference on Acoustics, Speech and Signal Processing (ICASSP), pp. 3305–3308. IEEE (2012). https://doi.org/10.1109/ICASSP.2012.6288622
17. Mazumdar, A.: Nonadaptive group testing with random set of defectives. IEEE Trans. Inf. Theory **62**(12), 7522–7531 (2016). https://doi.org/10.1109/TIT.2016.2613870
18. Scarlett, J., Cevher, V.: Near-optimal noisy group testing via separate decoding of items. In: 2018 IEEE International Symposium on Information Theory (ISIT), pp. 2311–2315. IEEE (2018). https://doi.org/10.1109/ISIT.2018.8437667
19. Scarlett, J., Johnson, O.: Noisy non-Adaptive group testing: a (near-)definite defectives approach. IEEE Trans. Inf. Theory **66**(6), 3775–3797 (2020). https://doi.org/10.1109/TIT.2020.2970184
20. Sejdinovic, D., Johnson, O.: Note on noisy group testing: asymptotic bounds and belief propagation reconstruction. In: 2010 48th Annual Allerton Conference on Communication, Control, and Computing (Allerton), pp. 998–1003. IEEE (2010). https://doi.org/10.1109/ALLERTON.2010.5707018
21. Sham, P., Bader, J.S., Craig, I., O'Donovan, M., Owen, M.: DNA pooling: a tool for large-scale association studies. Nat. Rev. Genet. **3**(11), 862–871 (2002). https://doi.org/10.1038/nrg930
22. Shani-Narkiss, H., Gilday, O.D., Yayon, N., Landau, I.D.: Efficient and practical sample pooling for high-throughput PCR diagnosis of COVID-19 (2020). https://www.medrxiv.org/content/10.1101/2020.04.06.20052159v2. Accessed 14 Jan 2021
23. Shental, N., et al.: Efficient high throughput SARS-CoV-2 testing to detect asymptomatic carriers. Sci. Adv. **6**(37), (2020). https://doi.org/10.1126/sciadv.abc5961
24. Täufer, M.: Rapid, large-scale, and effective detection of COVID-19 via non-adaptive testing. J. Theor. Biol. **506**(110450) (2020). https://doi.org/10.1016/j.jtbi.2020.110450, http://www.sciencedirect.com/science/article/pii/S0022519320303052
25. Wu, S., Wei, S., Wang, Y., Vaidyanathan, R., Yuan, J.: Achievable partition information rate over noisy multi-access Boolean channel. In: IEEE International Symposium on Information Theory - Proceedings, pp. 1206–1210. IEEE (2014). https://doi.org/10.1109/ISIT.2014.6875024

Phylogenetics

Novel Phylogenetic Network Distances Based on Cherry Picking

Kaari Landry[1(✉)], Aivee Teodocio[1], Manuel Lafond[2],
and Olivier Tremblay-Savard[1]

[1] Department of Computer Science, University of Manitoba, 66 Chancellors Cir,
Winnipeg, MB, Canada
landryk1@cs.umanitoba.ca
[2] Département d'informatique, Université de Sherbrooke,
2500 Boulevard de l'Université, Sherbrooke, QC, Canada

Abstract. In phylogenetic networks, picking a cherry consists of removing a leaf that shares a parent with another leaf, or removing a reticulate edge whose endpoints are parents of leaves. Cherry-picking operations were recently shown to have several structural and algorithmic applications in the study of networks, for instance in determining their reconstructibility or in solving the network hybridization and network containment problems. In particular, some classes of networks are isomorphic if they can be reduced to a single vertex by the same sequence of cherry-picking operations. Therefore, cherry-picking sequences contain information on the level of similarity between two networks. In this paper, we expand on this idea by devising four novel distances on networks based on cherry picking and their reverse operation. We provide bounds between these distances and show that three of them are equal despite their different formulations. We also show that computing these three equivalent distances is NP-hard, even when restricted to comparing a tree and a network. On the positive side, we show that they can be computed in quadratic time on two trees, providing a new comparative measure for phylogenetic trees that can be computed efficiently.

Keywords: Graphs and networks · Trees · Network problems ·
Applications · Algorithm design and analysis · Biology and genetics

1 Introduction

Biology research has long been concerned with establishing evolutionary relationships between species. The first illustrations that were drawn of these relationships took the form of "trees of life" [5,8,17]. Today this work is mostly accomplished in the realm of bioinformatics and is based on the analysis of genomic sequences instead of physical traits to produce these phylogenies. Phylogenomics has been mainly focused on constructing trees since those early days, but these types of graphs are only able to represent vertical relationships, in the

© Springer Nature Switzerland AG 2021
C. Martín-Vide et al. (Eds.): AlCoB 2021, LNBI 12715, pp. 57–81, 2021.
https://doi.org/10.1007/978-3-030-74432-8_5

direction of ancestor to descendant. Hybridization events, such as the ones created by cross-pollination or crossbreeding in plants [6,26], and horizontal gene transfer events, which are more common in prokaryotes [16,24] but are believed to be occurring in eukaryotes as well [9,15], are best represented by phylogenetic networks instead. Furthermore, the recombination of viruses is well-known and network-like [7]. This is also of special concern to human health, where recombination is considered to be very important to the evolution of HIV and contributes to the virus's fast adaptation and diversification [25]. Phylogenetic networks introduce the concept of reticulations [22], which allow internal nodes to have multiple parents, and thus represent the transfer of information from other species than the direct ancestor. With recent research focusing more on these evolutionary relationships, phylogenetic networks are gaining more traction as a better way of illustrating links between species.

Since errors in the phylogenetic construction pipeline, missing data, the use of different datasets or different construction methods/models can result in incompatible evolutionary relationships [1], distance methods allowing to compare phylogenies are necessary to measure differences between them. Distance calculations can also be used to evaluate the performance of a new construction approach (by comparing a phylogeny produced by a new tool with a state-of-the-art manually curated one) and to evaluate the differences between a gene tree and a species tree. In the context of phylogenetic trees, the Robinson-Foulds distance [21] is by far the most widely used metric, mostly because of its simplicity and ease of calculation.

Cardona *et al.* presented in 2008 a generalization of the Robinson-Foulds distance that is computed in polynomial time for tree-child time-consistent networks [2]. This distance is based on calculating the symmetric difference between the sets of all clusters (a cluster is the set of leaves descendant of a vertex) corresponding to the two compared networks. Lu *et al.* [19] developed a fast exponential-time algorithm for computing the soft Robinson-Foulds distance between phylogenetic networks, which compares soft clusters between networks, *i.e.* clusters that appear in each of the trees represented by the network.

Finding new ways to characterize phylogenetic trees and networks can also lead to new distance definitions. In 2013, Humphries *et al.* [10] presented a new type of characterization for phylogenetic trees named *cherry-picking sequences*, which was later reformulated in 2019 by co-authors Linz and Semple [18]. Essentially, a cherry-picking move corresponds to the removal of a leaf that is part of a cherry (*i.e.* two sibling leaves connected to the same parent), and a cherry-picking sequence (CPS) represents a series of cherry-picking moves that can reduce a phylogenetic tree to a single leaf. These sequences were initially defined to help characterize the hybridization number for a set of phylogenies, *i.e.* the minimum number of reticulations required to build a network that represents all the phylogenies. Since the introduction of cherry-picking sequences, they have also been used to solve other problems, such as the network hybridization problem [12] and the network containment problem [13,14]. Interestingly, it was shown in [14] that defining an ordering on the cherry-picking sequences allows to find a unique smallest CPS to represent a reconstructible cherry-picking network (*i.e.* a network that can be reduced by a CPS and reconstructed by the same CPS). Moreover, it follows that two cherry-picking networks within the same reconstructible class are isomorphic if and only if they have the same smallest CPS [14]. Since cherry-picking operations can be used to decide whether two cherry-picking networks are identical, one may ask whether they can also be used to infer a level of similarity or dissimilarity between two networks. This fact is what motivated us to explore how distances between phylogenetic trees and networks could be defined using the concept of cherry-picking sequences.

In this paper, we begin by presenting several different distances based on cherry-picking sequences. We first propose novel operational distances based on the minimum number of cherry operations required to transform one network into the other, or that are required to make the networks reach a common substructure. We also introduce the tail distance, which is based on the structure of the networks' CPSs and asks for a CPS on each network that maximizes a common suffix. We show that three of our proposed distances are equivalent, but that they are NP-hard to compute even when comparing a network and a tree. On the positive side, we show that these distances can be computed in polynomial time when comparing two trees, providing a novel measure between phylogenetic trees. To develop our NP-hardness reductions, we also introduce the CP-SUBTREE problem, which asks whether a tree can be obtained from a network by applying cherry-picking operations, and prove that it is NP-hard. This problem may be of independent interest since it represents a new form of tree containment, a widely studied problem in the area of phylogenetic networks.

2 Preliminaries

2.1 Network Definitions

Given a directed graph $D = (V, E)$, we write $d^+(v)$ and $d^-(v)$ for the number of out-neighbors and in-neighbors, respectively, of a node $v \in V$. We may write $V(D)$ and $E(D)$ to refer to the vertices and edges of D, respectively. A phylogenetic X-*network*, $\mathcal{N} = (V, E, X)$, is a rooted acyclic directed graph where the leaves are bijectively labeled by a set of taxa X. We refer to the leaves of $\mathcal{N}$ as $L(\mathcal{N})$ and assume that $d^+(l) = 0$ and $d^-(l) = 1$ for each $l \in L(\mathcal{N})$. For simplicity, a phylogenetic X-network is referred to as hereafter as a network, with the understanding that $X = L(\mathcal{N})$. Edges are directed to the sinks $L(\mathcal{N})$ from the root $r(\mathcal{N})$, which is the unique vertex with $d^-(r(\mathcal{N})) = 0$ and $d^+(r(\mathcal{N})) \geq 2$. An internal node $v \in V \setminus L(\mathcal{N})$ is a *tree node* when $d^-(v) = 1$ and $d^+(v) \geq 2$, and v is a *reticulation node* when $d^-(v) \geq 2$ and $d^+(v) = 1$. We write $T(\mathcal{N})$ and $R(\mathcal{N})$ for the set of tree nodes and reticulation nodes, respectively. Unless otherwise stated, we assume that trees and networks are *binary*, that is, for a tree or network $\mathcal{N}$, $d^+(r(\mathcal{N})) = 2$, $d^+(v) = 2$ for all $v \in T(\mathcal{N})$, and $d^-(v) = 2$ for all $v \in R(\mathcal{N})$. It may be that $|V(\mathcal{N})| = |L(\mathcal{N})| = 1$, in which case, we call $\mathcal{N}$ a *single-leaf network*. For a network $\mathcal{N}$, $V(\mathcal{N}) = \{r(\mathcal{N})\} \cup T(\mathcal{N}) \cup R(\mathcal{N}) \cup L(\mathcal{N})$.

For an edge $(u, v) \in E(\mathcal{N})$ with $v \in T(\mathcal{N}) \cup L(\mathcal{N})$, we say that u is the *parent* of v and v is a child of u, denoted $p(v) = u$. Generally, if there is a path from u to v, then u is an *ancestor* of v and v is a *descendant* of u. A node v is *unary* if $d^-(v) = d^+(v) = 1$. *Suppressing* a unary node consists of adding an edge from $p(v)$ to the child of v, and removing v with its incident edges. We write $\mathcal{N} \simeq \mathcal{N}'$ if $\mathcal{N}$ and $\mathcal{N}'$ are isomorphic networks, with preservation of the leaf labels.

2.2   Cherry Operations

In phylogenetic trees, a cherry usually refers to two leaves that have the same parent. In networks, the authors of [14] proposed to distinguish two types of cherries. Let x, y be two distinct leaves of a network $\mathcal{N}$. We say that (x, y) is a *non-reticulated cherry* if $p(x) = p(y)$. Note that in this case, (y, x) refers to the same cherry, although the order will be relevant later on. Analogously, we say that (x, y) is a *reticulated cherry* if $p(x) \in R(\mathcal{N})$ and $(p(y), p(x)) \in E$. Note that in this case, $p(y) \notin R(\mathcal{N})$ and (y, x) is not a cherry. We let $\mathcal{C}_c(\mathcal{N})$ and $\mathcal{C}_r(\mathcal{N})$ denote the set of non-reticulated and reticulated cherries of a network $\mathcal{N}$, respectively, and denote $\mathcal{C}(\mathcal{N}) = \mathcal{C}_c(\mathcal{N}) \cup \mathcal{C}_r(\mathcal{N})$.

A sequence of cherries $S = \langle (x_1, y_1), \ldots, (x_n, y_n) \rangle$ is called a *cherry sequence* (CS) and is used to define a series of cherry operations. The reverse of S is denoted $rev(S) = \langle (x_n, y_n), \ldots, (x_1, y_1) \rangle$. The concatenation of two CSs S_1 and S_2 is denoted $S_1 \cdot S_2$. CSs have hitherto in the literature been referred to as a cherry-picking sequence (CPS), but here we prefer a more general term, since CSs will refer to either network reduction operations, or network expansion operations.

Cherry Reduction. A *cherry reduction*, or *cherry picking*, is a cherry operation that modifies a network by deleting a cherry of either type. To be more specific, let $\mathcal{N}$ be a network. Then a cherry reduction on $\mathcal{N}$ of (x, y) consists of one of the following, depending on the cherry type of (x, y):

- If $(x, y) \in \mathcal{C}_c(\mathcal{N})$, then remove x and its incident edge from $\mathcal{N}$. If $p(x)$ is not the root of $\mathcal{N}$, then suppress $p(x)$. If $p(x)$ is the root of $\mathcal{N}$, then remove $p(x)$ and its incident edge.
- If $(x, y) \in \mathcal{C}_r(\mathcal{N})$, then remove the edge $(p(y), p(x))$ and suppress the two resulting unary nodes.
- If $(x, y) \notin \mathcal{C}(\mathcal{N})$, then the cherry reduction of (x, y) has no effect on $\mathcal{N}$. Such a reduction will be called *trivial*, and *non-trivial* otherwise.

Figure 1 illustrates both cases of non-trivial cherry reductions (from (a) to (b) and from (b) to (c)).

For a CS S, $\mathcal{N}\langle S \rangle$ denotes the network that results from reducing the network $\mathcal{N}$ by the cherries of S in order. Assume that a cherry sequence S reduces $\mathcal{N}$ to a single-leaf network. Then

$$|S| \geq \frac{|V(\mathcal{N})| - 1}{2} \tag{1}$$

noting that every cherry reduction removes two vertices and binary networks always have an odd number of vertices.

Not every network can be reduced to a single-leaf network by a CS of cherry reductions, since there exist networks with no cherry. A network that can be reduced to a single-leaf network by a sequence of cherry reductions is called a *cherry-picking network* (CPN) or *orchard*. One example of a known network relationship that prevents cherry picking is when two reticulations are exclusive siblings. For a leaf child of a reticulation to be a part of a cherry, the reticulation must have a leaf sibling, which this structure prevents in the binary context. From here on all networks referred to are assumed to be a CPN unless otherwise specified.

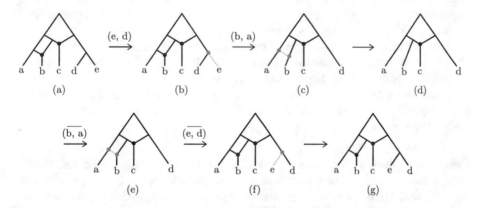

Fig. 1. Networks (a) through (g) have reticulated nodes indicated in black. Vertices and edges targeted by an operation are shown in grey and dotted. Networks (a) through (d) show steps in the reduction of the network in (a) by a non-reticulated cherry (e,d) and a reticulated cherry (b,a). Networks (d) through (g) illustrate the cherry expansion that reverses the previous reduction operations, giving the network in (g) that is isomorphic to the network in (a).

Cherry Expansion. A *cherry expansion* is an operation that adds a cherry leaf or a reticulation. More precisely, let $\mathcal{N}$ be a network. Then the cherry expansion on $\mathcal{N}$ of (x, y) consists of the following, depending on its type:

- If $x \notin L(\mathcal{N})$ and $y \in L(\mathcal{N})$, then check if $p(y)$ exists. If so, subdivide the edge $(p(y), y)$ creating the new vertex u to which the new leaf x is to be attached by a new edge (u, x). If $p(y)$ does not exist, then $\mathcal{N}$ is a single leaf network. In this case, create a new root with children x and y.
- If $x, y \in L(\mathcal{N})$, subdivide the edge $(p(y), y)$ creating a vertex u, and subdivide the edge $(p(x), x)$ creating a vertex v. Then insert the new edge (u, v) so that v becomes a reticulation.
- If $y \notin L(\mathcal{N})$ then (x, y) for any $x \in L(\mathcal{N})$ or $x \notin L(\mathcal{N})$, has no effect on $\mathcal{N}$. Such an expansion will be called *trivial*, and *non-trivial* otherwise.

For a CS S, we denote by $\mathcal{N}\langle\overline{S}\rangle$ the network obtained by applying each cherry expansion in S in order (the *overline* notation emphasizes that the pairs in S must be interpreted as expansions rather than reductions).

Figure 1(d) to (g) demonstrates both non-trivial cherry expansions.

So for a network $\mathcal{N}$ and a CS S of cherry expansions

$$|V(\mathcal{N}\langle\overline{S}\rangle)| \leq |V(\mathcal{N})| + 2|S| \tag{2}$$

The cherry expansion is the reverse operation to a cherry reduction, where the class of binary CPNs have been described by Janssen and Murakami as being a *reconstructible class* [14]. There are two corollaries to this feature important to our purposes. First, for a network $\mathcal{N}$ with a cherry $c \in \mathcal{C}(\mathcal{N})$, $\mathcal{N}\langle c\rangle\langle\overline{c}\rangle \simeq \mathcal{N}$ holds, which we refer to as the *reversibility of cherry-picking*. Second, from any single-leaf network $\mathcal{N}$ where $L(\mathcal{N}) = \{x\}$, every binary CPN containing x can be reached from $\mathcal{N}$ by some sequence of cherry expansions.

We will refer to a CS of cherry reductions as a *reducing CS*, and refer to a CS of cherry expansions as an *expanding CS*. It might sometimes be desirable for a CS to contain both types of operations, with the overline and no-overline notation being used on a cherry-by-cherry basis to indicate the intended operation. We call a CS of this kind a *mixed CS*. For instance in Fig. 1, the sequence from networks (a) to (g) is a mixed CS.

Complete Cherry Sequences and Cherry-Picking Subnetworks. For the remainder, we will assume that cherry operations in a CS S always affect the network and that S contains no trivial operation. Let S_i denote the sequence with the i first elements of S and let (x, y) be the $(i + 1)$-th element of S. If (x, y) is a cherry reduction, we assume that $(x, y) \in \mathcal{C}(\mathcal{N}\langle S_i\rangle)$, and if (x, y) is a cherry expansion, we assume $y \in L(\mathcal{N}\langle S_i\rangle)$. In this way, Eqs. 1 and 2 are always equalities. If S satisfies the above property, then S is called *minimal*.

If a reducing CS S is minimal and reduces a network $\mathcal{N}$ to a single-leaf network, then we say S is *complete* for $\mathcal{N}$. It is worth noting that a complete CS can always be found for a CPN. It is also known that applying a cherry reduction on a CPN results in another CPN [14, Observation 1].

CSs with non-trivial entries can be used to determine whether two networks are isomorphic. The next general-purpose result is analogous to Lemma 12 in [14], a result proving that the complete reduction of a binary network by the complete reducing CS for another network implies the former is a subnetwork of the latter. In the case where the complete reducing CS for both networks is the same, the lemma implies isomorphism between the networks by showing set equality, each input being a subnetwork of the other. The proof for the following lemma and all other lemmas and theorems are found in the appendix.

Lemma 1. *Let $\mathcal{N}_1$ and $\mathcal{N}_2$ be two networks with at least two leaves, and assume that there is a CS S that is complete for both $\mathcal{N}_1$ and $\mathcal{N}_2$. Then $\mathcal{N}_1 \simeq \mathcal{N}_2$.*

We finish this section by introducing cherry-picking subnetworks, which is the cherry-picking analogous notion of the well-studied agreement subtrees (see [23]) and agreement subnetworks (see [3]). Given two networks $\mathcal{N}$ and $\mathcal{N}'$, we say that $\mathcal{N}'$ is a *cherry-picking subnetwork* of $\mathcal{N}$ if there exists a reducing CS S such that $\mathcal{N}\langle S\rangle \simeq \mathcal{N}'$. If this is the case, we will write $\mathcal{N}' \subseteq_{cp} \mathcal{N}$. An *agreement cherry-picking subnetwork* (ACPS) of two networks $\mathcal{N}_1$ and $\mathcal{N}_2$ is a network $\mathcal{N}'$ satisfying $\mathcal{N}' \subseteq_{cp} \mathcal{N}_1$ and $\mathcal{N}' \subseteq_{cp} \mathcal{N}_2$. The *maximum agreement cherry-picking subnetwork* (MACPS) for networks $\mathcal{N}_1$ and $\mathcal{N}_2$ is an ACPS that maximizes the total number of vertices. As we shall see, MACPSs are closely related to cherry-picking distances, but it is NP-hard to decide whether $\mathcal{N}_1 \subseteq_{cp} \mathcal{N}_2$.

3   Distances Based on Cherry-Picking Sequences

This section introduces new network distances based on cherry operations and CSs. Hereafter, we assume that $\mathcal{N}_1$ and $\mathcal{N}_2$ are cherry-picking networks, not necessarily on the same set of leaves. We do assume however that $L(\mathcal{N}_1) \cap L(\mathcal{N}_2) \neq \emptyset$, as otherwise all distances are ∞. Similarly, if one of $\mathcal{N}_1$ or $\mathcal{N}_2$ is not a cherry-picking network, then all the distances are defined to be ∞.

1. The *Mixed distance*, $d_m(\mathcal{N}_1, \mathcal{N}_2)$, for networks $\mathcal{N}_1$ and $\mathcal{N}_2$, is the length of a minimum mixed CS that transforms network $\mathcal{N}_1$ into $\mathcal{N}_2$ when its cherry operations are applied in order.
2. The *Construction distance*, $d_c(\mathcal{N}_1, \mathcal{N}_2)$, for networks $\mathcal{N}_1$ and $\mathcal{N}_2$, is the minimum combined length of a reducing CS S^- and an expanding CS S^+ where $\mathcal{N}_1$ reaches $\mathcal{N}_2$ when reduced by S^- then expanded by S^+. In other words, $d_c(\mathcal{N}_1, \mathcal{N}_2)$ is the minimum of $|S^-| + |S^+|$ over all CS pairs S^- and S^+ satisfying $\mathcal{N}_1\langle S^-\rangle\langle \overline{S^+}\rangle \simeq \mathcal{N}_2$.
3. The *Deconstruction distance*, $d_d(\mathcal{N}_1, \mathcal{N}_2)$, for networks $\mathcal{N}_1$ and $\mathcal{N}_2$, is the minimum of $|S_1| + |S_2|$ over all CS pairs S_1 and S_2 satisfying $\mathcal{N}_1\langle S_1\rangle \simeq \mathcal{N}_2\langle S_2\rangle$.
4. The *Tail distance*, $d_{tail}(\mathcal{N}_1, \mathcal{N}_2)$, for networks $\mathcal{N}_1$ and $\mathcal{N}_2$, is the minimum combined length of reducing CSs S_1 and S_2 that can both be concatenated with a common CS S to produce complete CSs for $\mathcal{N}_1$ and $\mathcal{N}_2$. In other words, $d_{tail}(\mathcal{N}_1, \mathcal{N}_2)$ is the minimum of $|S_1| + |S_2|$ over all pairs of *complete* CSs $S_1 \cdot S$ and $S_2 \cdot S$ for $\mathcal{N}_1$ and $\mathcal{N}_2$, respectively.

Intuitively speaking, d_m minimizes the number of operations required to turn $\mathcal{N}_1$ into $\mathcal{N}_2$, whereas d_c requires first applying reductions, followed by expansions. On the other hand, d_d minimizes the number of reductions required on both networks to reach a common subnetwork and d_{tail} focuses on finding complete sequences with the maximum number of common operations in the suffix.

In the rest of this section, we show how each of these distances are related. Particularly, we show that d_c, d_d and d_{tail} are equal, but not always equal to d_m.

Theorem 2. *For any networks $\mathcal{N}_1$ and $\mathcal{N}_2$, $d_m(\mathcal{N}_1, \mathcal{N}_2) \leq d_c(\mathcal{N}_1, \mathcal{N}_2)$.*

To further elaborate on the finding of Theorem 2, Fig. 2 shows that $d_m(\mathcal{N}_1, \mathcal{N}_2) < d_c(\mathcal{N}_1, \mathcal{N}_2)$ is possible. As illustrated, the difference between these distances can be arbitrarily large, so d_c can hardly be used to approximate d_m.

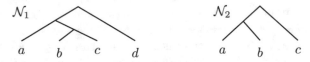

Fig. 2. Two networks $\mathcal{N}_1$ (left) and $\mathcal{N}_2$ (right), which happen to be trees. One minimum mixed CS that can be used to determine $d_m(\mathcal{N}_1, \mathcal{N}_2)$ is $(c, b)\overline{(c, d)}(d, c)$. One example of two (combined) minimum CSs that can be used to determine $d_c(\mathcal{N}_1, \mathcal{N}_2)$ is the reducing CS $(c, b)(b, a)(d, a)$ and the expanding CS $(c, a)(b, a)$. Note what happens to the CS lengths if we attach more descendant leaves to the edge $(p(a), a)$ (in any binary configuration): the combination giving $d_c(\mathcal{N}_1, \mathcal{N}_2)$ can be made arbitrarily large while $d_m(\mathcal{N}_1, \mathcal{N}_2)$ stays constant.

Theorem 3. *For any networks $\mathcal{N}_1$ and $\mathcal{N}_2$, $d_c(\mathcal{N}_1, \mathcal{N}_2) = d_d(\mathcal{N}_1, \mathcal{N}_2)$.*

Theorem 4. *For any networks $\mathcal{N}_1$ and $\mathcal{N}_2$, $d_{tail}(\mathcal{N}_1, \mathcal{N}_2) = d_d(\mathcal{N}_1, \mathcal{N}_2)$.*

It has now been shown for networks $\mathcal{N}_1$ and $\mathcal{N}_2$ that $d_{tail}(\mathcal{N}_1, \mathcal{N}_2)$, $d_d(\mathcal{N}_1, \mathcal{N}_2)$, and $d_c(\mathcal{N}_1, \mathcal{N}_2)$ are equivalent and bounded from below by $d_m(\mathcal{N}_1, \mathcal{N}_2)$. We finish this section by establishing a relationship between these distances and agreement cherry-picking subnetworks.

Theorem 5. *Let $\mathcal{N}_1 = (V_1, E_1)$ and $\mathcal{N}_2 = (V_2, E_2)$ be networks and let $\mathcal{N}^* = (V^*, E^*)$ be a MACPS of $\mathcal{N}_1$ and $\mathcal{N}_2$. Then, $d_d(\mathcal{N}_1, \mathcal{N}_2) = \frac{|V_1|-1}{2} + \frac{|V_2|-1}{2} - |V^*| + 1$.*

4   Computing the Tail Distance Between Two Trees

In this section, we show that computing $d_{tail}(T_1, T_2)$ between two trees T_1 and T_2 can be done in polynomial time, even if $L(T_1) \neq L(T_2)$. This is achieved by showing that computing d_{tail} is equivalent to finding a special type of maximum agreement subtree, which can be found in time $O(|V(T_1)||V(T_2)|)$ using dynamic programming. This polynomial time solvability does not extend as we later show the problem to be NP-hard for a network-tree comparison.

A *tree* is a network with no reticulations. Let T be a tree. For $x, y \in L(T)$, $lca_T(x, y)$ is the lowest common ancestor of x and y, which is well-defined. Let $S \subseteq L(T)$. By $T|S$, we denote the smallest connected subgraph of T that contains each leaf in S. Note that $T|S$ may contain unary nodes. By $T||S$, we denote the tree obtained from $T|S$ after suppressing unary nodes.

Focusing on a certain restriction, for non-empty $S \subseteq L(T)$, we say that $T||S$ is a *leaf-restricted subtree* of T if either $|S| = 1$, or if both of the following hold:

1. There exist $x, y \in S$ such that $lca_T(x, y) = r(T)$.
2. If v is a unary node of $T|S$, then every descendant of v in $T|S$ is either a unary node or a leaf.

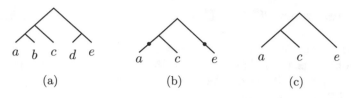

(a)   (b)   (c)

Fig. 3. (a) A tree T in (a) on $X = \{a, b, c, d, e\}$. (b) The restriction $T|S$, where $S = \{a, c, e\}$ with unary nodes marked. (c) The restriction $T\|S$, noting that this tree is a leaf-restricted subtree of the tree in (a).

We write $T' \subseteq_{lr} T$ if T' is isomorphic to $T\|L(T')$ and if $T\|L(T')$ is a leaf-restricted subtree of T. Roughly speaking, unless $|L(T')| = 1$, to have $T' \subseteq_{lr} T$, Condition 1 requires $r(T)$ to be part of T', and Condition 2 requires that all removed vertices to be lying between a leaf and its parent (see Fig. 3). Having $T' \subseteq_{lr} T$ implies T' is *displayed* by T (T' can be obtained from T through edge and vertex deletions, up to suppressing unary vertices), in addition to satisfying these additional conditions. This lets us establish the connection with cherry reductions.

Given two phylogenetic trees T_1 and T_2, a *leaf-restricted agreement subtree* between T_1 and T_2 is a tree T^* such that both $T^* \subseteq_{lr} T_1$ and $T^* \subseteq_{lr} T_2$ hold. A *leaf-restricted maximum agreement subtree*, or LR-MAST for short, is a leaf-restricted agreement subtree with the maximum number of leaves. Note that since an LR-MAST T^* is a tree, $|V(T^*)| = 2|L(T^*)| - 1$, so maximizing the number of vertices is the same as maximizing the number of leaves.

We first prove the transitivity of the $\subseteq_{lr}$ relationship.

Lemma 6. *If $T_1 \subseteq_{lr} T_2$ and $T_2 \subseteq_{lr} T_3$, then $T_1 \subseteq_{lr} T_3$.*

We can now connect $\subseteq_{lr}$ with cherry sequences, as it turns out to be equivalent to the $\subseteq_{cp}$ relationship.

Lemma 7. *Let T be a tree. Then $T' \subseteq_{lr} T$ if and only if $T' \subseteq_{cp} T$.*

Next, we show that d_{tail} can be computed easily if an LR-MAST T^* is known.

Theorem 8. *Let T_1, T_2 be two trees, and let T^* be an LR-MAST of T_1 and T_2. Then $d_{tail}(T_1, T_2) = (|V(T_1)| + |V(T_2)|)/2 - |V(T^*)|$.*

4.1 Computing an LR-MAST

Because maximizing $|V(T^*)|$ is equivalent to maximizing $|L(T^*)|$, Theorem 8 shows that, on trees, computing $d_{tail}(T_1, T_2)$ is equivalent to computing an LR-MAST. This can be achieved using a dynamic programming procedure. This standard dynamic programming algorithm is described in Appendix 4 which also provides a theorem and proof that lead to the following.

Theorem 9. *For two given trees T_1 and T_2, $d_{tail}(T_1, T_2)$, and therefore $d_c(T_1, T_2)$ and $d_d(T_1, T_2)$, can be computed in time $O(|V(T_1)||V(T_2)|)$.*

5　NP-hardness of d_{tail} Between a Tree and a Network

We show that computing $d_{tail}(T, N)$ between a tree T and a network N is NP-hard, holding even when $L(T) \subseteq L(N)$. In fact, we show something stronger, deciding $T \subseteq_{cp} N$ is NP-hard. This directly implies computing d_{tail} is NP-hard. The CP-SUBTREE problem.

Input: a tree T and a network N such that $L(T) \subseteq L(N)$.

Question: is $T \subseteq_{cp} N$?

Roughly speaking, the hardness of CP-SUBTREE implies the hardness of computing d_{tail} since, given a network N and a tree T, the smallest d_{tail} one can hope for is $|V(N) - V(T)|/2$. This is because we need to at least make N and T have the same number of vertices, and each cherry reduction affects two vertices. Furthermore, we can attain this distance if and only if $T \subseteq_{cp} N$, establishing the equivalence between the two problems.

Now we show the CP-SUBTREE is NP-hard with a reduction from the 3-SAT-3-OCC problem, known to be NP-hard (see [20] and [4, Problem 1]). This version of 3-SAT gives sets of clauses, each of which contains either 2 or 3 literals related by logical *or* denoted $\vee$. Each literal occurs exactly three times, twice positively and once negatively (a literal is a boolean variable x_i or its negation $\overline{x}_i$). Examples of clauses are $C_1 = (x_1 \vee \overline{x}_2)$ or $C_2 = (\overline{x}_1 \vee \overline{x}_2 \vee \overline{x}_3)$. The goal is to decide if an assignment of the x_i variables exists satisfying every clause.

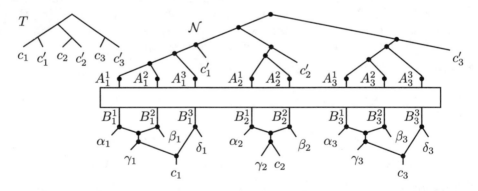

Fig. 4. The main structure of T and N obtained from an instance with three clauses C_1, C_2, C_3. Here, N is obtained from T by replacing each leaf c_i by a substructure. In this example, C_1 and C_3 have 3 literals whereas C_2 has 2 literals. Not shown: children of the A_i^j vertices and parents of the B_i^j vertices, which make the connections between clauses and variables.

Let ϕ be a 3-SAT-3-OCC instance, with n variables $x_1, \ldots, x_n$ and m clauses $C_1, \ldots, C_m$. We construct a corresponding tree T and network N. Let T be any tree with $2m$ leaves and m cherries, each cherry representing a clause C_i. Name the cherry's leaves that correspond to C_i by c_i and c_i'. Obtain a network N by starting at T, then replacing each leaf c_i by a substructure as in Fig. 4.

More precisely, from T, replace each leaf c_i as follows, depending on the number of literals of clause C_i (we will specify how the A_i^j vertices are connected to the B_i^j vertices later):

- If C_i has 2 literals, replace leaf c_i by a tree with 2 leaves A_i^1, A_i^2. Add vertices B_i^1, B_i^2 and leaves $\alpha_i, \beta_i, \gamma_i, c_i$ as in the middle substructure in Fig. 4.
- If C_i has 3 literals, replace leaf c_i by a tree with 3 leaves A_i^1, A_i^2, A_i^3. Add vertices B_i^1, B_i^2, B_i^3 and leaves $\alpha_i, \beta_i, \gamma_i, \delta_i, c_i$ as in the left or right substructure in Fig. 4.

Conceptually, A_i^1, B_i^1 represent the first literal that can satisfy clause C_i, A_i^2, B_i^2 represent the second literal and A_i^3, B_i^3 the third literal (if present).

We can now specify the connections between the A_i^j's and B_i^j's. Consider a variable x_h and recall that it has two positive occurrences and one negative. Assume that x_h occurs positively in clauses C_i and C_j and negatively in clause C_k. Then there are two pairs of vertices A_i^a, B_i^a and A_j^b, B_j^b representing the positive occurrences x_h, and one pair of vertices A_k^c, B_k^c representing the negative occurrence $\overline{x}_h$. Add vertices and edges between these vertices as in Fig. 5, thereby creating two new leaves p_i^a and p_k^c. We will refer to the subgraph illustrated in Fig. 5 as the x_h gadget. We construct this gadget for each variable x_h. After this, since each A_i^a, B_i^a pair represents a distinct literal, each A_i^a vertex has exactly two children and each B_i^a vertex has exactly one parent.

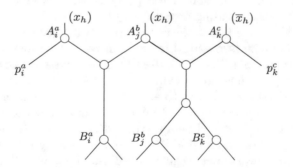

Fig. 5. The x_h gadget, exhibiting the relationship between the A and B vertices corresponding to the occurrences of x_h. Here, A_i^a and B_i^a are associated with choosing x_h to satisfy C_i; A_j^b, B_j^b are associated with choosing x_h to satisfy C_j; and A_k^c, B_k^c are associated with choosing $\overline{x}_h$ to satisfy C_k.

This concludes the construction. The main intuition is that we need to get rid of dummy leaves to turn N into T. To achieve this, each c_i leaf will need to be "brought up" to its c_i' sibling by eliminating the substructures of N. We must thus choose a path for each c_i, either through A_i^1, A_i^2 or A_i^3 (if present). Such a choice will correspond to choosing a literal to satisfy each C_i, and the x_h gadgets prevent us from choosing two paths/literals that are contradictory.

Theorem 10. *The CP-SUBTREE problem is NP-hard.*

Theorem 11. *The problem of deciding whether $d_{tail}(T, N) \leq k$ for given network N, tree T and integer k is NP-hard.*

Corollary 12. *The problem of computing the distances d_{tail}, d_c or d_d on two given networks is NP-hard.*

6   Discussion and Future Work

These novel distances have certain advantages and disadvantages, we can also reflect on their algorithmic complexity, and also on their context in network classes.

In terms of accuracy, one immediately obvious feature of these distances is that they have a "bottom bias". That is, disagreements in the networks closer to the leaves are easier to access with cherry operations than those disagreements closer to the root. A specific result of this bias is that the tail distance won't penalize misplaced leaves near other leaves. However, differences in the networks have a gradually increasing cost with the further the occurrence is from a cherry.

In this work we have only provided a single algorithm for the case of the inputs being two trees. In fact, a naive algorithm to calculate these distances on networks may run in exponential time. Future work in engineering an algorithm for calculating these distances will consider the possibility of approximation, heuristics, and parameterization of the problem on a variable other than the input size, such as the resulting distance.

Then there is the question of the context of network classes. Our findings rely on the results from Janssen and Murakami on reconstructible network classes, four of which are defined. Here we looked at the reconstructible class of binary CPNs, which begs the question of whether our results apply to the other three reconstructible classes. If we want to extend our work to general networks, this would require some assurance that the network can be reduced to a single leaf. It has been shown by Linz and Semple that attaching new (non-X) leaves can transform any network into a CPN [18]. Extending cherry operation-based distances to general networks may require a leaf-attaching operation.

It is also interesting to note that boundaries of the CPN class were not well defined, other than by cherry operations, until recently. Work from van Iersel et al. provide a structural characterization of the CPN class for the first time [11]. It would be interesting to see how the cherry operation-based distances may be simplified or not by their definition of cherry covers.

Appendix

1   Proof of Lemma 1 (Page 6)

Lemma 1. *Let $\mathcal{N}_1$ and $\mathcal{N}_2$ be two networks with at least two leaves, and assume that there is a CS S that is complete for both $\mathcal{N}_1$ and $\mathcal{N}_2$. Then $\mathcal{N}_1 \simeq \mathcal{N}_2$.*

Proof. We prove the statement by induction on $|S|$. As a base case, assume that $|S| = 1$. Then $S = ((x,y))$ for some x, y and, since $\mathcal{N}_1\langle S\rangle$ and $N_2\langle S\rangle$ are both single-leaf networks and (x,y) is non-trivial, it must be that $\mathcal{N}_1$ and $\mathcal{N}_2$ both consist of a tree with two leaves x and y. Hence $\mathcal{N}_1 \simeq \mathcal{N}_2$.

For the induction step, suppose that $|S| \geq 2$. Let (x,y) be the first element of S and let S' be obtained from S by removing (x,y) (so that $S = ((x,y)) \cdot S'$). Let $\mathcal{N}_1' = \mathcal{N}_1\langle(x,y)\rangle$ and $\mathcal{N}_2' = \mathcal{N}_2\langle(x,y)\rangle$. Notice that S' is complete for both $\mathcal{N}_1'$ and $\mathcal{N}_2'$ since it reduces them to a single-leaf network and does not contain trivial reductions (as otherwise, S would contain trivial reductions). Thus by induction, $\mathcal{N}_1' \simeq \mathcal{N}_2'$. Now consider the (x,y) reduction. First assume that $(x,y) \in \mathcal{C}_c(\mathcal{N}_1)$. Then x is removed from $\mathcal{N}_1$ when (x,y) is applied to $\mathcal{N}_1$. Since $\mathcal{N}_1' \simeq \mathcal{N}_2'$, x cannot be a leaf of $\mathcal{N}_2'$, and since (x,y) is non-trivial with respect to $\mathcal{N}_2$, it must be that x is removed from $\mathcal{N}_2$ by the (x,y) reduction. That is, $(x,y) \in \mathcal{C}_c(\mathcal{N}_2)$ as well. It follows that $\mathcal{N}_1 \simeq \mathcal{N}_2$ because both networks are obtained from $\mathcal{N}_1'$ and $\mathcal{N}_2'$ by adding the x leaf back on the branch from y to its parent.

Assume instead that $(x,y) \in C_r(x,y)$. Then when applied on $\mathcal{N}_1$, (x,y) removes $(p(y), p(x))$ from $\mathcal{N}_1$ and thus $x \in L(\mathcal{N}_1')$. Therefore, because $\mathcal{N}_1' \simeq \mathcal{N}_2'$, applying (x,y) cannot remove x from $L(\mathcal{N}_2)$ and, since (x,y) is non-trivial with respect to $\mathcal{N}_2$, it must be that $(x,y) \in \mathcal{C}_r(\mathcal{N}_2)$ as well. It follows that $\mathcal{N}_1 \simeq \mathcal{N}_2$ because both networks are obtained from $\mathcal{N}_1'$ and $\mathcal{N}_2'$ by adding an edge from the branch above y to the branch above x.

2 Proof of Theorems in Sect. 3 (Theorem 2 Page 7, Theroem 3 Page 8, Theorem 4 Page 8, Theorem 5 Page 8)

Theorem 2. *For any networks $\mathcal{N}_1$ and $\mathcal{N}_2$, $d_m(\mathcal{N}_1, \mathcal{N}_2) \leq d_c(\mathcal{N}_1, \mathcal{N}_2)$.*

Proof. Let S^- and S^+ be CSs satisfying $\mathcal{N}_1\langle S^-\rangle\langle \overline{S^+}\rangle \simeq \mathcal{N}_2$ such that $d_c(\mathcal{N}_1, \mathcal{N}_2) = |S^-| + |S^+|$. Then $S' = S^- \cdot S^+$ is a mixed CS and applying S' on $\mathcal{N}_1$ transforms it into $\mathcal{N}_2$, implying $d_m(\mathcal{N}_1, \mathcal{N}_2) \leq d_c(\mathcal{N}_1, \mathcal{N}_2)$.

Theorem 3. *For any networks $\mathcal{N}_1$ and $\mathcal{N}_2$, $d_c(\mathcal{N}_1, \mathcal{N}_2) = d_d(\mathcal{N}_1, \mathcal{N}_2)$.*

Proof. For the first direction we show that $d_d(\mathcal{N}_1, \mathcal{N}_2) \leq d_c(\mathcal{N}_1, \mathcal{N}_2)$. Let S^- be a reducing CS and S^+ be an expanding CS that minimize $|S^-| + |S^+|$ such that $\mathcal{N}_1\langle S^-\rangle\langle \overline{S^+}\rangle \simeq \mathcal{N}_2$. Thus $d_c(\mathcal{N}_1, \mathcal{N}_2) = |S^-| + |S^+|$. By the reversibility of cherry picking $\mathcal{N}_1\langle S^-\rangle \simeq \mathcal{N}_2\langle rev(S^+)\rangle$. It follows that $d_d(\mathcal{N}_1, \mathcal{N}_2) \leq |S^-| + |rev(S^+)| = |S^-| + |S^+| = d_c(\mathcal{N}_1, \mathcal{N}_2)$.

For the second direction we show that $d_c(\mathcal{N}_1, \mathcal{N}_2) \leq d_d(\mathcal{N}_1, \mathcal{N}_2)$. Let S_1 and S_2 be reducing CSs that minimize $|S_1| + |S_2|$ such that $\mathcal{N}_1\langle S_1\rangle \simeq \mathcal{N}_2\langle S_2\rangle$. Thus $d_d(\mathcal{N}_1, \mathcal{N}_2) = |S_1| + |S_2|$. By reversibility of cherry picking, $\mathcal{N}_1\langle S_1\rangle\langle \overline{rev(S_2)}\rangle \simeq \mathcal{N}_2$. It follows that $d_c(\mathcal{N}_1, \mathcal{N}_2) \leq |S_1| + |rev(S_2)| = |S_1| + |S_2| = d_d(\mathcal{N}_1, \mathcal{N}_2)$.

Theorem 4. *For any networks $\mathcal{N}_1$ and $\mathcal{N}_2$, $d_{tail}(\mathcal{N}_1, \mathcal{N}_2) = d_d(\mathcal{N}_1, \mathcal{N}_2)$.*

Proof. We first show that $d_d(\mathcal{N}_1, \mathcal{N}_2) \leq d_{tail}(\mathcal{N}_1, \mathcal{N}_2)$. Let S_1 (respectively S_2) be a reducing CS that is complete for $\mathcal{N}_1$ (respectively $\mathcal{N}_2$), such that $S_1 = S_1' \cdot S$ and $S_2 = S_2' \cdot S$ for some S of maximum size. Thus $d_{tail}(\mathcal{N}_1, \mathcal{N}_2) = |S_1'| + |S_2'|$. Since S_1 and S_2 are complete, we know that S reduces $\mathcal{N}_1\langle S_1'\rangle$ and $\mathcal{N}_2\langle S_2'\rangle$ to a single-leaf network and does not contain trivial entries. By Lemma 1, $\mathcal{N}_1\langle S_1'\rangle \simeq \mathcal{N}_2\langle S_2'\rangle$. Since S_1' and S_2' are possible CSs for d_d, it follows that $d_d(\mathcal{N}_1, \mathcal{N}_2) \leq |S_1'| + |S_2'| = d_{tail}(\mathcal{N}_1, \mathcal{N}_2)$.

We next show that $d_{tail}(\mathcal{N}_1, \mathcal{N}_2) \leq d_d(\mathcal{N}_1, \mathcal{N}_2)$. Let S_1, S_2 be CSs such that $\mathcal{N}_1\langle S_1\rangle \simeq \mathcal{N}_2\langle S_2\rangle$ and such that $|S_1| + |S_2|$ is minimum. Then $d_d(\mathcal{N}_1, \mathcal{N}_2) = |S_1| + |S_2|$. Define $\mathcal{N}^* = \mathcal{N}_1\langle S_1\rangle$ (which is a CPN and is isomorphic to $\mathcal{N}_2\langle S_2\rangle$) and let S be a complete reducing CS for $\mathcal{N}^*$. Then $\mathcal{N}_1\langle S_1 \cdot S\rangle$ is a single-leaf network and $\mathcal{N}_1\langle S_2 \cdot S\rangle$ is also a single-leaf network. Since S is a common tail, it follows that $d_{tail}(\mathcal{N}_1, \mathcal{N}_2) \leq |S_1| + |S_2| = d_d(\mathcal{N}_1, \mathcal{N}_2)$.

Theorem 5. *For any networks $\mathcal{N}_1$ and $\mathcal{N}_2$, $d_{tail}(\mathcal{N}_1, \mathcal{N}_2) = d_d(\mathcal{N}_1, \mathcal{N}_2)$.*

Proof. Let S_1 and S_2 be reducing CSs such that $\mathcal{N}_1\langle S_1\rangle \simeq \mathcal{N}^*$ and $\mathcal{N}_2\langle S_2\rangle \simeq \mathcal{N}^*$. It is clear that $d_d(\mathcal{N}_1, \mathcal{N}_2) \leq |S_1| + |S_2|$. We show that equality holds. Assume for contradiction that $d_d(\mathcal{N}_1, \mathcal{N}_2) < |S_1| + |S_2|$. Then there are S_1', S_2' such that $\mathcal{N}_1\langle S_1'\rangle \simeq \mathcal{N}_2\langle S_2'\rangle$. Thus $\mathcal{N}_1\langle S_1'\rangle$ is an ACPS of $\mathcal{N}_1$ and $\mathcal{N}_2$. However, since $|S_1'| + |S_2'| < |S_1| + |S_2|$, we must have $|S_1'| < |S_1|$ or $|S_2'| < |S_2|$ (or both). Assume without loss of generality that $|S_1'| < |S_1|$. Then $\mathcal{N}_1\langle S_1'\rangle$ has strictly more vertices than $\mathcal{N}^* = \mathcal{N}_1\langle S_1\rangle$ since less vertices need to be removed. This contradicts the fact that $\mathcal{N}^*$ is a MACPS. Therefore, $d_d(\mathcal{N}_1, \mathcal{N}_2) = |S_1| + |S_2|$.

Now, let S^* be a complete CS for $\mathcal{N}^*$. Then

$$d_d(\mathcal{N}_1, \mathcal{N}_2) = |S_1| + |S_2| = (|S_1| + |S^*|) - |S^*| + (|S_2| + |S^*|) - |S^*|$$

$$= \frac{|V_1| - 1}{2} - \frac{|V^*| - 1}{2} + \frac{|V_2| - 1}{2} - \frac{|V^*| - 1}{2} \quad \text{by Eq. 1}$$

$$= \frac{|V_1| - 1}{2} + \frac{|V_2| - 1}{2} - |V^*| + 1$$

3 Proof of Lemma 6, 7 and Theorem 8 (Page 9)

Lemma 6. *If $T_1 \subseteq_{lr} T_2$ and $T_2 \subseteq_{lr} T_3$, then $T_1 \subseteq_{lr} T_3$.*

Proof. For the duration of the proof, denote $L_1 = L(T_1)$, $L_2 = L(T_2)$ and $L_3 = L(T_3)$. First notice that by the conditions of the lemma, $L_1 \subseteq L_2 \subseteq L_3$. Since T_2 is isomorphic to $T_3||L_2$, it is not hard to see that for any $X \subseteq L_2$, $T_2||X$ is isomorphic to $T_3||X$. In particular, $T_2||L_1$ is isomorphic to $T_3||L_1$, which in turn is isomorphic to T_1 (because $T_1 \subseteq_{lr} T_2$, and thus $T_1 \simeq T_2||L_1$). It only remains to show that $T_3||L_1$ satisfies both conditions of leaf-restricted subtrees.

If $|L_1| = 1$, then this is trivially the case, so suppose that $|L_1| \geq 2$. Since $T_2||L_1$ is a leaf-restricted subtree of T_2, there must be $x, y \in L_1$ such that $lca_{T_2}(x, y) = r(T_2)$. In a similar vein, since $T_2 \subseteq_{lr} T_3$, there are $x', y' \in L_2$

such that $lca_{T_3}(x', y') = r(T_3)$. Moreover, because T_2 is isomorphic to $T_3 \| L_2$, $lca_{T_2}(x', y') = r(T_2)$ must hold. The fact that $lca_{T_2}(x, y) = lca_{T_2}(x', y')$ then implies that $lca_{T_3}(x, y) = r(T_3)$, and so $T_3 \| L_1$ satisfies Condition 1.

Now assume that $T_3 \| L_1$ does not satisfy Condition 2. That is, there is a unary node v in $T_3 | L_1$ that has a descendant w with two children. Let v be the deepest node of $T_3 | L_1$ with this property. Then we may assume that v is the parent of w in $T_3 | L_1$ (and thus in T_3). Note that because $T_3 \| L_1$ satisfies Condition 1, v is on the path from $r(T_3)$ to w. Let a, b be leaves of $T_3 | L_1$ such that $lca_{T_3 | L_1}(a, b) = w$.

Since $L_1 \subseteq L_2$, a, b, w and v are all in $T_3 | L_2$. However, v is not unary in $T_3 | L_2$ because $T_2 \subseteq_{lr} T_3$. Therefore, there exists $d \in L_2$ such that $lca_{T_3}(d, a) = v$. Let $w' = lca_{T_2}(a, b)$ and $v' = lca_{T_2}(d, a)$. Since T_2 is isomorphic to $T_3 \| L_2$ and v is the parent of w in T_3, v' is the parent of w' in T_2. But then, v' must be unary in $T_2 | L_1$, since otherwise, there would exist $d' \in L_1$ such that $lca_{T_2 | L_1}(d', a)$ is the parent of $lca_{T_2 | L_1}(a, b)$ in $T_2 | L_1$. If this was the case, v would not be unary in $T_3 | L_1$. Also, w' is not unary in $T_2 | L_1$ since $a, b \in L_1$. The existence of w' and v' contradict $T_1 \subseteq_{lr} T_2$. We deduce that $T_3 \| L_1$ satisfies Condition 2, and therefore that $T_1 \subseteq_{lr} T_3$.

Lemma 7. *Let T be a tree. Then $T' \subseteq_{lr} T$ if and only if $T' \subseteq_{cp} T$.*

Proof. We first note that if T' has only one node, it consists of a leaf of T. In that case, $T' \subseteq_{lr} T$ holds, and one can easily find C such that $T\langle C \rangle = T'$. Thus we assume that T', and thus T, has more than one leaf, and prove the two directions of the lemma separately.

($\Rightarrow$) We first show that if $T' \subseteq_{lr} T$, then $T' = T\langle C \rangle$ for some cherry sequence C. We prove the statement by induction over the quantity $k := |L(T)| - |L(T')|$. As a base case, assume that $k = 0$. Then T' is isomorphic to T and the empty cherry-picking sequence proves this case.

Now, consider the inductive step with $k > 0$. For convenience, denote $S = L(T')$, noting that T' is isomorphic to $T \| S$. Assume that in T, there is a cherry $c = (x, y)$ such that $x \notin S$. Because $T\langle c \rangle$ is obtained by deleting x and suppressing $p(x)$, it is not hard to see that $T \| S$ is isomorphic to $T\langle c \rangle \| S$, and thus that $T' \subseteq_{lr} T\langle c \rangle$. Furthermore, we have $|L(T\langle c \rangle)| - |L(T')| = k - 1$, and we know by induction that there is a cherry-picking sequence transforming $T\langle c \rangle$ into T', hence a cherry-picking from T to T'.

So suppose that no cherry c as described above exists. That is, for every cherry (x, y) of T, both $x, y \in S$. We know that $S \neq L(T)$ since $k > 0$, so there exists some leaf $x \in V(T) \setminus S$. By our assumption, x cannot belong to a cherry of T, and so the child w of $p(x)$ other than x must be an internal node of T. Thus, there is a cherry (y, z) in T, in the subtree induced by w and its descendants. By our assumption, $y, z \in S$, and it follows that the parent w' of y and z in T must be in $T | S$. But then, by the first condition of leaf-restricted subtrees, $r(T)$ must be in $T | S$. Moreover, $p(x)$ must be in $T | S$ since it lies on the path between the root and w'. However, $p(x)$ has only one child in $T | S$ since $x \notin S$, and $p(x)$ has descendant w' with two children in $T | S$. This contradicts the second condition of the definition of $T' \subseteq_{lr} T$, and so this case does not occur. We deduce that there does exist a CS from T to T'.

($\Leftarrow$) We now show that if there exists a CS $C = (c_1, \dots, c_k)$ such that $T' \simeq T\langle C\rangle$, then $T\langle C\rangle \subseteq_{lr} T$. In fact, by Lemma 6, it suffices to show that if T' is obtained from T after applying one cherry reduction, then $T' \subseteq_{lr} T$. The statement will follow, since if T_i denotes the i-th tree after applying the i-th cherry reduction of C, this shows that $T_k \subseteq_{lr} T_{k-1} \subseteq_{lr} \dots \subseteq_{lr} T_1 \subseteq_{lr} T$, and then $T_k \subseteq_{lr} T$ by transitivity (Lemma 6).

Therefore, assume that $T' = T\langle c\rangle$ for some cherry $c = (x, y)$. Let g be the parent of $p(x)$. Then from T to T', we remove x and suppress its parent $p(x)$. Defining $S = L(T) \setminus \{x\}$, it is then easy to see that $T\langle c\rangle = T||S$. Moreover, x could not be a child of the root since otherwise, c would not be a cherry (recall that $|L(T')| > 1$). This implies that $r(T)$ is in $T||S$, satisfying Condition 1. Moreover, only $p(x)$ can be unary in $T|S$ and Condition 2 is satisfied since $p(x)$ has two children leaves. Therefore, $T\langle c\rangle \subseteq_{lr} T$, as desired.

Theorem 8. *Let T_1, T_2 be two trees, and let T^* be an LR-MAST of T_1 and T_2. Then $d_{tail}(T_1, T_2) = (|V(T_1)| + |V(T_2)|)/2 - |V(T^*)|$.*

Proof. Let us denote $n_1 := |V(T_1)|$, $n_2 = |V(T_2)|$ and $n^* = |V(T^*)|$. Let $C_1 = (a_1, \dots, a_k)$ and $C_2 = (b_1, \dots, b_h)$ be cherry-picking sequences to apply on T_1 and T_2, respectively, so that $T_1\langle C_1\rangle$ and $T_2\langle C_2\rangle$ are isomorphic, and such that $k + h$ is minimum among all possibilities. By Theorem 4, we know that $d_{tail}(T_1, T_2) = k + h$. Furthermore, by Lemma 7, we know that $T_1\langle C_1\rangle \subseteq_{lr} T$ and $T_2\langle C_2\rangle \subseteq_{lr} T$, and thus $T_1\langle C_1\rangle$ is a leaf-restricted agreement subtree of T_1 and T_2. Denote $\hat{n} = |V(T_1\langle C_1\rangle)|$ (which is equal to $|V(T_2\langle C_2\rangle)|$). Notice that $\hat{n} \leq n^*$ because T^* maximizes the number of nodes among all leaf-restricted agreement subtrees. Also note that each cherry reduction removes two nodes, namely a leaf and its parent. Therefore, $T_1\langle C_1\rangle$ has $n_1 - 2k$ nodes and $T_2\langle C_2\rangle$ has $n_2 - 2h$ nodes, i.e. $\hat{n} = n_1 - 2k = n_2 - 2h$. Isolating k and h, it follows that

$$d_{tail}(T_1, T_2) = k + h = (n_1 - \hat{n})/2 + (n_2 - \hat{n})/2$$
$$= (n_1 + n_2)/2 - \hat{n}$$
$$\geq (n_1 + n_2)/2 - n^*$$

This proves the first bound of the proof. Consider now the complementary bound. By Lemma 7, since $T^* \subseteq_{lr} T_1$ and $T^* \subseteq_{lr} T_2$, there is a cherry sequence C_1 from T_1 to T^*, and another cherry sequence C_2 from T_2 to T^*. Because a cherry reduction removes two nodes, C_1 must contain $(n_1 - n^*)/2$ elements, and C_2 must contain $(n_2 - n^*)/2$ elements. Since $T_1\langle C_1\rangle$ and $T_2\langle C_2\rangle$ are isomorphic, it follows that $d_{tail}(T_1, T_2) \leq (n_1 - n^*)/2 + (n_2 - n^*)/2 = (n_1 + n_2)/2 - n^*$. This concludes the proof.

4   LR-MAST Algorithm, and Associated Theorem (Page 9)

For a tree T and $v \in V(T)$, $T[v]$ denotes the subtree induced by v and all of its descendants. We may write $L(v)$ instead of $L(T[v])$.

For $u \in V(T_1)$ and $v \in V(T_2)$, $M[u, v]$ denotes the number of leaves in an LR-MAST between $T_1[u]$ and $T_2[v]$. The idea is to combine the LR-MAST on both sides of u and v and combine them into a larger LR-MAST, when possible. This is similar to the classical MAST algorithm, but with less cases to check because of the additional restrictions. We now describe the recurrence for M.

Leaf Case. If u or v is a leaf, then $M[u, v] = 1$ if $L(u) \cap L(v) \neq \emptyset$, and $M[u, v] = 0$ otherwise.

Internal Node Case. Assume that u and v are both internal nodes. Let u_1, u_2 be the children of u, and v_1, v_2 be the children of v. We compute two temporary values $M_1[u, v]$ and $M_2[u, v]$, and take the maximum of the two. More precisely, apply the following:

$$
M_1[u, v] = \begin{cases}
0 & \text{if } L(u_1) \cap L(v_1) = \emptyset \text{ and } L(u_2) \cap L(v_2) = \emptyset \\
1 & \text{if } L(u_1) \cap L(v_1) \neq \emptyset \text{ and } L(u_2) \cap L(v_2) = \emptyset \\
1 & \text{if } L(u_1) \cap L(v_1) = \emptyset \text{ and } L(u_2) \cap L(v_2) \neq \emptyset \\
M[u_1, v_1] + M[u_2, v_2] & \text{if } L(u_1) \cap L(v_1) \neq \emptyset \text{ and } L(u_2) \cap L(v_2) \neq \emptyset
\end{cases}
$$

$$
M_2[u, v] = \begin{cases}
0 & \text{if } L(u_1) \cap L(v_2) = \emptyset \text{ and } L(u_2) \cap L(v_1) = \emptyset \\
1 & \text{if } L(u_1) \cap L(v_2) \neq \emptyset \text{ and } L(u_2) \cap L(v_1) = \emptyset \\
1 & \text{if } L(u_1) \cap L(v_2) = \emptyset \text{ and } L(u_2) \cap L(v_1) \neq \emptyset \\
M[u_1, v_2] + M[u_2, v_1] & \text{if } L(u_1) \cap L(v_2) \neq \emptyset \text{ and } L(u_2) \cap L(v_1) \neq \emptyset
\end{cases}
$$

and put

$$M[u, v] = \max(M_1[u, v], M_2[u, v])$$

Roughly speaking, M_1 corresponds to mapping u_1 to v_1 and u_2 to v_2, and M_2 corresponds to mapping u_1 to v_2 and u_2 to v_1. We now show that these recurrences are correct.

Theorem 13. $M[u, v]$ *as defined above is the number of leaves in an LR-MAST between T_1 and T_2.*

Proof. We proceed by induction over the height of the trees $T_1[u]$ and $T_2[v]$. Assume that one of u or v is a leaf, say u without loss of generality. If $L(u) \cap L(v) \neq \emptyset$, then $u \in L(v)$ and the tree consisting of only u is an LR-MAST, hence $M[u, v] = 1$. Otherwise, there is no leaf in common, there cannot exist an LR-MAST, and $M[u, v] = 0$ is correct.

Now assume that u and v are internal nodes. Let T^* be an LR-MAST between $T_1[u]$ and $T_2[v]$, and let S be such that T^* is isomorphic to $T_1[u]|||S$ and $T_2[v]|||S$. First assume that $S = \emptyset$, i.e. T^* has no vertex. Then it must be that $L(u) \cap L(v) = \emptyset$. In that case, by induction $M[u_1, v_1] = M[u_1, v_2] = M[u_2, v_1] = M[u_2, v_2] = 0$ since none of the subtrees below u and v can share a leaf either. Thus $M[u, v] = 0$ will be correctly set.

Assume now that $|S| = 1$. Then $L(u) \cap L(v) \geq 1$. Thus one of $L(u_1) \cap L(v_1), L(u_1) \cap L(v_2), L(u_2) \cap L(v_1)$ or $L(u_2) \cap L(v_2)$ is non-empty, implying that one or both of $M_1[u, v]$ or $M_2[u, v]$ is at least 1. Moreover, one of $L(u_1) \cap L(v_1)$ and $L(u_2) \cap L(v_2)$ must be empty, as otherwise a leaf-restricted agreement subtree with two leaves can be found, contradicting $|S| = 1$. Similarly, one of $L(u_1) \cap L(v_2)$ and $L(u_2) \cap L(v_1)$ must be empty. It follows that $M_1[u, v] \leq 1$ and $M_2[u, v] \leq 1$, and that $\max(M_1[u, v], M_2[u, v]) = 1$. Thus $M[u, v] = 1$ will be correctly set.

Now assume that $|S| \geq 2$. We first argue that $M[u, v] \geq |L(T^*)|$. Let $U_1 = L(u_1) \cap S$ and $U_2 = L(u_2) \cap S$. Similarly, let $V_1 = L(v_1) \cap S$ and $V_2 = L(v_2) \cap S$. By the Condition 1 of leaf-restricted subtrees, S must contain leaves a, b such that $lca_{T_1[u]}(a, b) = u$, and leaves a', b' such that $lca_{T_2[v]}(a', b') = v$. This implies that none of U_1, U_2, V_1, V_2 is empty. We have that $T_1[u]|||S$ contains u and $T_2[v]|||S$ contains v, and since both subtrees are isomorphic, it follows that either:

(1) $T_1[u_1]|||U_1$ is isomorhpic to $T_2[v_1]|||V_1$ and $T_1[u_2]|||U_2$ is isomorphic to $T_2[v_2]|||V_2$; or that
(2) $T_1[u_1]|||U_1$ is isomorphic to $T_2[v_2]|||V_2$ and $T_1[u_2]|||U_2$ is isomorphic to $T_2[v_1]|||V_1$.

Assume that (1) holds. Then $L(u_1) \cap L(v_1) \neq \emptyset$ and $L(u_2) \cap L(v_2) \neq \emptyset$, so the $M_1[u, v]$ entry will be set by the fourth case in the recurrence. We now argue that $T_1[u_1]|||U_1$ is a leaf-restricted agreement subtree of $T_1[u_1]$. If $|U_1| = 1$, this trivially holds, so assume $|U_1| \geq 2$. We prove that both conditions of leaf-restricted subtrees hold:

- Condition 1. If there are no $a, b \in U_1$ such that $lca_{T_1[u_1]}(a, b) = u_1$, then all members of U_1 are below one child of u_1. This means that in $T_1[u]|||S$, u_1 is unary, but has a descendant of degree 2 since $|U_1| \geq 2$, a contradiction to the fact that $T_1[u]|||S$ is a leaf-restricted subtree. Thus Condition 1 holds for $T_1[u_1]|||U_1$.
- Condition 2. There cannot be a unary node w with a binary descendant in $T_1[u_1]|||U_1$, as otherwise w would also be unary in $T_1[u]|||U_1$ and would also have a binary descendant.

It follows that $T_1[u_1]|||U_1$ is a leaf-restricted subtree of $T_1[u_1]$. By the same reasoning, $T_2[v_1]|||V_1$ is a leaf-restricted subtree of $T_2[v_1]$. And since $T_1[u_1]|||U_1$ and $T_2[v_1]|||V_1$ are isomorphic, they are both leaf-restricted agreement subtrees of $T_1[v_1]$ and $T_2[v_2]$. By symmetry, $T_1[u_2]|||U_2$ is a leaf-restricted agreement subtree of $T_1[u_2]$ and $T_2[v_2]$.

We do not know if $T_1[u_1]|||U_1$ and $T_1[u_2]|||U_2$ maximize the number of nodes among all agreement subtrees, but, by induction, we know that $M[u_1, v_1] \geq |U_1|$ and $M[u_2, v_2] \geq |U_2|$, and thus

$$M[u, v] \geq M[u_1, v_1] + M[u_2, v_2] \geq |U_1| + |U_2| = |L(T^*)|$$

If (2) holds, a similar reasoning shows that $M[u, v] \geq M[u_1, v_2] + M[u_2, v_1] \geq |U_1| + |U_2|$, and again $M[u, v] \geq |L(T^*)|$.

We now show that $M[u,v] \leq |L(T^*)|$. Suppose that $M_1[u,v] \geq M_2[u,v]$, so that $M[u,v] = M_1[u,v]$ (the case $M_2[u,v] \geq M_1[u,v]$ is identical). If $M_1[u,v] \leq 1$, then $M[u,v] = M_1[u,v] \leq 2 \leq |L(T^*)|$, as desired.

Thus assume $M_1[u,v] \geq 2$. This is only possible if $L(u_1) \cap L(v_1)$ and $L(u_2) \cap L(v_2)$ are non-empty. It follows that $M[u_1, v_1] > 0$ and $M[u_2, v_2] > 0$. By induction, there exist $A \subseteq L(u_1) \cap L(v_1)$ such that $T_1[u_1]|||A$ is an LR-MAST of $T_1[u_1]$ and $T_2[v_1]$ with $|A| = M[u_1, v_1]$ leaves, and there exists $B \subseteq L(u_2) \cap L(v_2)$ such that $T_1[u_2]|||B$ is an LR-MAST of $T_1[u_2]$ and $T_2[v_2]$ with $|B| = M[u_2, v_2]$ leaves. To ease notation, write

$$T_A = T_1[u_1]|||A \quad \text{and} \quad T_B = T_1[u_2]|||B$$

Consider the tree T' obtained by joining the roots of T_A and T_B under a common parent. Note that T' has $M[u_1, v_1] + M[u_2, v_2]$ leaves. We want to argue that $T' \subseteq_{lr} T[u]$ and $T' \subseteq_{lr} T[v]$.

It is clear that T' is isomorphic to $T_1[u]|||(A \cup B)$ and to $T_2[v]|||(A \cup B)$. Thus it suffices to show that T' satisfies the conditions of leaf-restricted subtrees. Since T_A and T_B are non-empty, T' contains a leaf from both sides of u and v and Condition 1 of leaf-restricted subtrees is satisfied. Consider Condition 2. All the nodes of T' have the same number of children as in T_A or T_B, except the root of T' which is not present in either tree. This root has two children, so we could not have created a new unary node with a binary descendant. Therefore Condition 2 is also satisfied.

We deduce that T' is a leaf-restricted agreement subtree of $T_1[u]$ and $T_2[v]$, and thus $|L(T')| \leq |L(T^*)|$. We therefore have

$$M[u,v] = M[u_1, v_1] + M[u_2, v_2] = |L(T')| \leq |L(T^*)|$$

as desired.

The case $M_2[u,v] \geq M_1[u,v]$ can be handled in the same manner by symmetry.

We have therefore shown that $|L(T^*)| \leq M[u,v] \leq |L(T^*)|$, proving equality.

The value of interest here is $M[r(T_1), r(T_2)]$. It can be obtained by calculating $M[u,v]$ for each $u \in V(T_1)$ and each $v \in V(T_2)$ in a bottom-up fashion. Since each calculation of each $M[u,v]$ entry can be done in time $O(1)$, assuming the values of the descendants of u and v have been computed and stored.

5   Proof of Theorem 10 (Page 12) and Theorem 11 (Page 12)

Theorem 10. *The CP-SUBTREE problem is NP-hard.*

Proof. Let ϕ be an instance of 3-SAT-3-OCC. Let T and N be constructed as described above. For the remainder of the proof, a leaf in $L(N) \setminus L(T)$ will be called a *dummy* leaf. We show that ϕ is satisfiable if and only if there exists a cherry sequence C such that $N\langle C \rangle$ is isomorphic to T.

($\Rightarrow$) : suppose that ϕ is satisfiable by some assignment Q of the x_i variables. We show that there is a cherry sequence that transforms N into T. Consider a clause C_i. Then Q satisfies the j-th literal of C_i for some $j \in \{1, 2, 3\}$. If there are multiple choices for j, choose one arbitrarily. Depending on j, apply one of the cherry sequences appearing in Fig. 6. The figure assumes a clause C_i with three literals. If C_i has two literals, then δ_i and B_i^3 do not exist, c_i is a sibling of γ_i and we have $j \in \{1, 2\}$. For such a clause, the cases for $j = 1$ or $j = 2$ can be applied as in Fig. 6, ignoring the first (c_i, δ_i) move.

As a result, for each $i \in \{1, \dots, m\}$, B_i^j becomes replaced by c_i when Q satisfies the j-th literal of C_i. Moreover, each $B_i^{j'}, j' \neq j$, has been replaced by a dummy leaf.

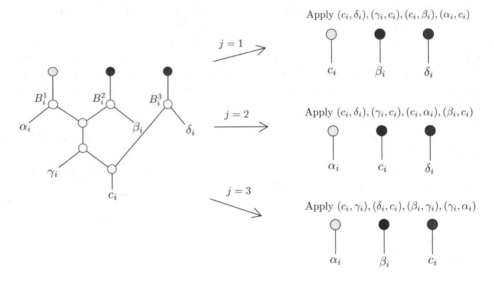

Fig. 6. The three possible ways to handle leaf c_i, depending on whether we satisfy clause C_i with its first, second or third literal, respectively corresponding to $j = 1, j = 2$ and $j = 3$. The colors on the nodes are there to emphasize which node receives c_i as a child after the cherry reduction sequence.

Next, consider a variable x_h. Assume that x_h occurs positively in clauses C_i and C_j and negatively in clause C_k. After applying the above cherry sequence for each clause, the x_h gadget has three leaves that have replaced the B_i^a, B_j^b and B_k^c vertices. From this point, we distinguish three possible cases:

(1) none of c_i, c_j or c_k is a leaf of the x_h gadget;
(2) c_i and/or c_j is a leaf of the x_h gadget. In this case, c_k cannot be a leaf of the x_h gadget, since this would imply that assignment Q assigns x_h positively and negatively.

(3) c_k is a leaf of the x_h gadget. In this case, none of c_i and c_j is a leaf of the x_h gadget because again, Q would otherwise assign x_h both positively and negatively.

The handling of cases (2) and (3) is shown in Fig. 7, top and bottom respectively.

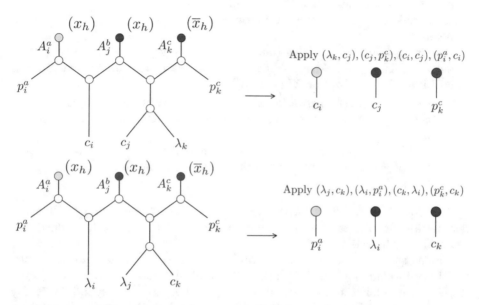

Fig. 7. Handling of Case 2 (top) and Case 3 (bottom) for the x_h gadget. Note that $\lambda_i, \lambda_j, \lambda_k$ are used to denote dummy leaves.

The handling of Case 1 can be performed as in Case 2, except that c_i and c_j are dummy leaves instead. The same applies in Case 2 when one of c_i or c_j is not present.

After applying this to every x_h gadget, each A_i^j vertex of N is replaced by either c_i or some dummy leaf. Importantly, notice that in all cases, every c_i leaf ends up replacing an A_i^j vertex, $j \in \{1, 2, 3\}$, and not some A_i^j leaf with $k \neq i$. In other words, N has become a tree such that for every $i \in \{1, \dots, m\}$, each of A_i^1, A_i^2, and A_i^3 if present, has become a leaf, one of which is c_i. It follows that at this point, N has become a tree identical to T but with extra dummy leaves. It is then easy to see that these can be removed by cherry-picking moves to obtain T. Therefore, $T \subseteq_{cp} N$.

($\Leftarrow$) : suppose that there exists a cherry sequence C that transforms N into T. We build a satisfying assignment Q for ϕ. The assignment is constructed by adding literals to Q—our goal is to add one literal per clause, without adding two contradictory literals x_h and $\overline{x}_h$.

Consider a c_i leaf of N, and let L_i be the literals that can satisfy clause C_i. If C_i has 2 literals, let $L_i = \{\ell_1, \ell_2\}$, and if C_i has 3 literals, let $L_i = \{\ell_1, \ell_2, \ell_3\}$.

Let X be the first cherry-picking move of C that contains c_i. We consider all possible cases.

1. If C_i has 2 literals, then c_i is a sibling of γ_i. In that case, go to Case 3 below.
2. If C_i has 3 literals, then either $X = (c_i, \gamma_i)$ or $X = (c_i, \delta_i)$. If $X = (c_i, \gamma_i)$, then add ℓ_3 to Q. In this case, notice that the next cherry-picking affecting c_i must delete δ_i, and that B_i^3 gets replaced by c_i.
   If instead $X = (c_i, \delta_i)$, then c_i becomes a sibling of γ_i. Then go to Case 3 below.
3. If c_i is a sibling of γ_i, notice that the next cherry-picking affecting c_i must delete γ_i. After γ_i is deleted, consider the next cherry-picking X' that contains c_i. Either $X' = (c_i, \alpha_i)$ or $X' = (c_i, \beta_i)$.
   - If $X' = (c_i, \alpha_i)$, then add ℓ_2 to Q. In this case, notice that the next cherry-picking of C affecting c_i must delete β_i, and that B_i^2 gets replaced by c_i.
   - If $X' = (c_i, \beta_i)$, then add ℓ_1 to Q. In this case, notice that the next cherry-picking of C affecting c_i must delete α_i, and that B_i^1 gets replaced by c_i.

It follows that in all cases, we add a literal ℓ_j to Q satisfying C_i. Thus Q satisfies all the clauses of ϕ. It remains to show that we have not added two contradictory literals x_h and $\overline{x}_h$.

Let us assume that we did add both x_h and $\overline{x}_h$. Assume that x_h occurs positively in clauses C_i and C_j and negatively in clause C_k. Let A_i^a, B_i^a and A_j^b, B_j^b correspond to the occurrences of x_h, and A_k^c, B_k^c correspond to $\overline{x}_h$. By our construction of Q, we know that $\overline{x}_h$ was added to Q because c_k ended up replacing B_k^c. Moreover, since x_h was added to Q, either c_i replaced B_i^a or c_j replaced B_j^b (or both). Now, notice that the next cherry reduction affecting c_k must be with the leaf that eventually replaces B_j^b, by the construction of the x_h gadget. Therefore, c_j cannot end up replacing B_j^b, since otherwise the cherry sequence C would have to remove one of c_j or c_k, preventing it from transforming N into T. Therefore, we may assume that x_h was added to Q because c_i ended up replacing B_i^a.

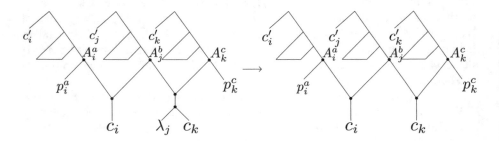

Fig. 8. The x_h gadget when C_i gets satisfied by $\overline{x}_h$ and C_k by x_h.

This situation is illustrated in Fig. 8. Note that this is just one possible situation, since we do not know when exactly c_i replaces B_i^a and when c_k replaces B_k^c, but our arguments hold only under the assumption that eventually, c_i replaces B_i^a and eventually, c_k replaces B_k^c.

We argue that after c_i replaces B_i^a, (c_i, p_i^a) cannot be in the cherry sequence C. If this were the case, (c_i, p_i^a) would remove the edge from A_i^a leading to c_i, and there would be no path from the parent of c_i' to c_i. This makes it impossible to reach T from that point since c_i and c_i' are siblings in T. In other words, we can never remove the unique edge from A_i^a that leads to c_i. By a similar argument, because we know that c_k replaces B_k^c eventually, we can never remove the unique edge from A_k^c that leads to c_k. It follows that at some point, we must reach the situation in Fig. 8 on the right. But then, the only two possible moves that involve one of c_i or c_k are (c_i, p_i^a) and (c_k, p_k^c). As we argued, both of these prevent us from reaching T, which is a contradiction.

We conclude that we could not have added two contradictory literals in Q, and thus ϕ is satisfiable.

Theorem 11. *The problem of deciding whether $d_{tail}(T, N) \leq k$ for given network N, tree T and integer k is NP-hard.*

Proof. We show that there is a polynomial-time reduction from CP-SUBTREE to the problem of computing d_{tail}. Let N, T be an instance of CP-SUBTREE. Our instance of computing d_{tail} is also N, T, and we put $k = |V(N) - V(T)|/2$. We show that $T \subseteq_{cp} N$ if and only if $d_{tail}(N, T) = \frac{|V(N)-V(T)|}{2}$.

($\Rightarrow$) Assume $T \subseteq_{cp} N$. First, note that the tail distance is equivalent to the deconstruction distance by Theorem 4.

This direction is easy to see in light of Theorem 5. The assumption $T \subseteq_{cp} N$ implies that T is a MACPS of N and T. This is because T is the largest possible agreement subnetwork of T and N that can be reached by cherry picking. It follows from Theorem 5 that

$$d_{tail}(N, T) = d_d(N, T) = \frac{|V(N)| - 1}{2} + \frac{|V(T)| - 1}{2} - |V(T)| + 1$$
$$= \frac{|V(N)| - |V(T)|}{2}$$

($\Leftarrow$) Assume that $d_{tail}(N, T) = \frac{|V(N)-V(T)|}{2}$. Let S^T be a complete CS for T, and let CS S^N be a complete CS for N. Say S^T and S^N have a maximum common ending subsequence S, so that

$$d_{tail}(N, T) = |S^N| + |S^T| - 2|S|$$

Substituting the given value of k for $d_{tail}(N, T)$ we get

$$\frac{|V(N)| - |V(T)|}{2} = |S^N| + |S^T| - 2|S|$$

We also know from Eq. 1 the length of a complete CS, so we substitute for the value of S^N.

$$\frac{|V(N)|}{2} - \frac{|V(T)|}{2} = \frac{|V(N)|}{2} - \frac{1}{2} + |S^T| - 2|S|$$

$$2|S| = |S^T| + \frac{|V(T)|}{2} - \frac{1}{2}$$

It happens to be that the last two terms on the right-hand side express the length of a complete reducing CS for T. Thus we have

$$2|S| = |S^T| + |S^T|$$
$$|S| = |S^T|$$

and S is a complete reducing sequence for T. If we write $S^N = S' \cdot S$, $N\langle S' \rangle$ yields a network for which S is a complete CS. By Lemma 1, $N\langle S' \rangle$ is isomorphic to T, implying that $T \subseteq_{cp} N$.

References

1. Anisimova, M., Liberles, D.A., Philippe, H., Provan, J., Pupko, T., von Haeseler, A.: State-of the art methodologies dictate new standards for phylogenetic analysis. BMC Evol. Biol. **13**, 161 (2013). https://doi.org/10.1186/1471-2148-13-161
2. Cardona, G., Llabrés, M., Rosselló, F., Valiente, G.: Metrics for phylogenetic networks I: Generalizations of the Robinson-foulds metric. IEEE/ACM Trans. Comput. Biol. Bioinf. **6**(1), 46–61 (2008)
3. Choy, C., Jansson, J., Sadakane, K., Sung, W.K.: Computing the maximum agreement of phylogenetic networks. Theoret. Comput. Sci. **335**(1), 93–107 (2005)
4. Condon, A., Maňuch, J., Thachuk, C.: The complexity of string partitioning. J. Discrete Algorithms **32**, 24–43 (2015)
5. Darwin, C.: On the origin of species by means of natural selection. Murray, London (1859). The preservation of favored races in the struggle for life
6. Ellstrand, N.C., Schierenbeck, K.A.: Hybridization as a stimulus for the evolution of invasiveness in plants? Proc. Nat. Acad. Sci. **97**(13), 7043–7050 (2000)
7. Froissart, R., Roze, D., Uzest, M., Galibert, L., Blanc, S., Michalakis, Y.: Recombination every day: abundant recombination in a virus during a single multi-cellular host infection. PLoS Biol **3**(3), e89 (2005)
8. Haeckel, E.: Generelle Morphologie der Organismen. Allgemeine Grundzüge der organischen Formen-Wissenschaft, mechanisch begründet durch die von C. Darwin reformirte Descendenz-Theorie, etc., vol. 2 (1866)
9. Hotopp, J.C.D.: Horizontal gene transfer between bacteria and animals. Trends Genet. **27**(4), 157–163 (2011)
10. Humphries, P.J., Linz, S., Semple, C.: Cherry picking: a characterization of the temporal hybridization number for a set of phylogenies. Bull. Math. Biol. **75**(10), 1879–1890 (2013). https://doi.org/10.1007/s11538-013-9874-x
11. van Iersel, L., Janssen, R., Jones, M., Murakami, Y., Zeh, N.: A unifying characterization of tree-based networks and orchard networks using cherry covers. arXiv preprint arXiv:2004.07677 (2020)
12. Janssen, Remie., Jones, Mark, Murakami, Yukihiro: Combining Networks Using Cherry Picking Sequences. In: Martín-Vide, Carlos, Vega-Rodríguez, Miguel A., Wheeler, Travis (eds.) AlCoB 2020. LNCS, vol. 12099, pp. 77–92. Springer, Cham (2020). https://doi.org/10.1007/978-3-030-42266-0_7

13. Janssen, Remie, Murakami, Yukihiro: Linear time algorithm for tree-child network containment. In: Martín-Vide, Carlos, Vega-Rodríguez, Miguel A., Wheeler, Travis (eds.) AlCoB 2020. LNCS, vol. 12099, pp. 93–107. Springer, Cham (2020). https://doi.org/10.1007/978-3-030-42266-0_8

14. Janssen, R., Murakami, Y.: On cherry-picking and network containment. arXiv preprint arXiv:1812.08065v2 (2020)

15. Keeling, P.J., Palmer, J.D.: Horizontal gene transfer in eukaryotic evolution. Nat. Rev. Genet. **9**(8), 605–618 (2008)

16. Koonin, E.V., Makarova, K.S., Aravind, L.: Horizontal gene transfer in prokaryotes: quantification and classification. Ann. Rev. Microbiol. **55**(1), 709–742 (2001)

17. de Lamarck, J.B.D.M.: Philosophie zoologique, ou Exposition des considérations relatives à l'histoire naturelle des animaux..., vol. 1. Dentu (1809)

18. Linz, S., Semple, C.: Attaching leaves and picking cherries to characterise the hybridisation number for a set of phylogenies. Adv. Appl. Math. **105**, 102–129 (2019)

19. Lu, B., Zhang, L., Leong, H.W.: A program to compute the soft Robinson-Foulds distance between phylogenetic networks. BMC Genomics **18**(2), 111 (2017). https://doi.org/10.1186/s12864-017-3500-5

20. Papadimitriou, C.: Computational Complexity. Addison-Wesley, Boston (1994)

21. Robinson, D.F., Foulds, L.R.: Comparison of phylogenetic trees. Math. Biosci. **53**(1–2), 131–147 (1981)

22. Sneath, P.H.: Cladistic representation of reticulate evolution. Syst. Zool. **24**(3), 360–368 (1975)

23. Steel, M., Warnow, T.: Kaikoura tree theorems: computing the maximum agreement subtree. Inf. Process. Lett. **48**(2), 77–82 (1993)

24. Thomas, C.M., Nielsen, K.M.: Mechanisms of, and barriers to, horizontal gene transfer between bacteria. Nat. Rev. Microbiol. **3**(9), 711–721 (2005)

25. Vuilleumier, S., Bonhoeffer, S.: Contribution of recombination to the evolutionary history of HIV. Curr. Opin. HIV AIDS **10**(2), 84–89 (2015)

26. Warwick, S.I., Stewart, C.: Crops come from wild plants: how domestication, transgenes, and linkage together shape ferality. Crop Ferality Volunteerism **36**(1), 9–30 (2005)

Best Match Graphs with Binary Trees

David Schaller[1], Manuela Geiß[2], Marc Hellmuth[3],
and Peter F. Stadler[1,4,5,6,7(✉)]

[1] Max Planck Institute for Mathematics in the Sciences, Leipzig, Germany
[2] Software Competence Center Hagenberg GmbH, Hagenberg, Austria
[3] Department of Mathematics, Faculty of Science, Stockholm University,
10691 Stockholm, Sweden
[4] Bioinformatics Group, Department of Computer Science, and Interdisciplinary
Center for Bioinformatics, Universität Leipzig, Härtelstrasse 16-18,
04107 Leipzig, Germany
studla@bioinf.uni-leipzig.de
[5] Institute for Theoretical Chemistry, University of Vienna, Vienna, Austria
[6] Facultad de Ciencias, Universidad Nacional de Colombia, Bogotá, Colombia
[7] Santa Fe Institute, Santa Fe, NM, USA

Abstract. Best match graphs (BMG) are a key intermediate in graph-based orthology detection and contain a large amount of information on the gene tree. We provide a near-cubic algorithm to determine whether a BMG can be explained by a fully resolved gene tree and, if so, to construct such a tree. Moreover, we show that all such binary trees are refinements of the unique binary-resolvable tree (BRT), which in general is a substantial refinement of the also unique least resolved tree of a BMG.

Keywords: Best match graphs · Binary trees · Rooted triple consistency · Polynomial-time algorithm

1 Introduction

The evolutionary history of a gene family can be described by a gene tree T, a species tree S, and an embedding of the gene tree into the species tree (Fig. 1A). The latter is usually formalized as a reconciliation map μ that locates gene duplication events along the edges of the species tree, identifies speciation events in T as those that map to vertices in S, and encodes horizontal gene transfer as edges in T that cross from one branch of S to another. Detailed gene family histories are a prerequisite for studying associations between genetic and phenotypic innovations. They also encode orthology, i.e., the notion that two genes in distinct species arose from a speciation event, which is a concept of key importance in genome annotation and phylogenomics. Two conceptually distinct approaches have been developed to infer orthology and/or complete gene family histories from sequence data. Tree-based methods explicitly construct the gene tree T and the species tree S, and then determine the reconciliation map μ as an optimization problem. Graph-based methods, on the other hand, start from *best*

C. Martín-Vide et al. (Eds.): AlCoB 2021, LNBI 12715, pp. 82–93, 2021.
https://doi.org/10.1007/978-3-030-74432-8_6

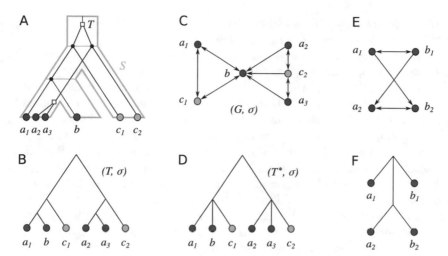

Fig. 1. (A) An evolutionary scenario consisting of a gene tree (T, σ) (whose topology is again shown in (B)) together with an embedding into a species tree S. The coloring σ of the leaves of T represents the species in which the genes reside. Speciation vertices ($\bullet$) of the gene tree coincide with the vertices of the species tree, whereas gene duplications ($\square$) are mapped to the edges of S. (C) The best match graph (G, σ) explained by (T, σ). (D) The unique least resolved tree (T^*, σ) explaining (G, σ). (E) An hourglass, i.e. the smallest example for a BMG that is not binary-explainable. (F) The (unique) tree that explains the hourglass.

matches, i.e., by identifying for each gene its closest relative or relatives in every other species. Due to the page limits, we only refer to a few key reviews and the references therein [12,13,21].

Best match graphs (BMGs) have only very recently been introduced as mathematical objects to formalize the idea of pairs of evolutionarily most closely related genes in two species [6]. The gene tree is modeled as a rooted, leaf-colored phylogenetic tree (T, σ) (Fig. 1B). Its leaf set $L(T)$ denotes the extant genes, and each gene $x \in L(T)$ is colored by the species $\sigma(x)$ in whose genome it resides. Given a tree T, we denote the *ancestor order* on its vertex set by $\preceq_T$. That is, we have $v \preceq_T u$ if u lies along the unique path connecting v to the root ρ_T of T, in which case we call u an *ancestor* of v. The *least common ancestor* $\text{lca}_T(A)$ is the unique $\preceq_T$-smallest vertex that is an ancestor of all genes in A. Writing $\text{lca}_T(x, y) := \text{lca}_T(\{x, y\})$, we have

Definition 1. *Let (T, σ) be a leaf-colored tree. A leaf $y \in L(T)$ is a* best match *of the leaf $x \in L(T)$ if $\sigma(x) \neq \sigma(y)$ and $\text{lca}_T(x, y) \preceq_T \text{lca}_T(x, y')$ holds for all leaves y' of color $\sigma(y') = \sigma(y)$.*

The *best match graph (BMG)* of a leaf-colored tree (T, σ), denoted by $G(T, \sigma)$, is a vertex-colored digraph with vertex set $L(T)$ and arcs (x, y) if and only if y is a best match of x (Fig. 1C). An arbitrary vertex-colored graph (G, σ) is a best match graph if there exists a leaf-colored tree (T, σ) such that $(G, \sigma) = G(T, \sigma)$. In this case, we say that (T, σ) *explains* (G, σ).

In practice, best matches (in the sense of Definition 1) are approximated by *best hits*, that is, pairs of genes with the largest sequence similarity. The two concepts are equivalent if a strict molecular clock is assumed. In the presence of rate variations, however, both best matches that are not best hits and best hits that not best matches may occur. We refer to [23] for a more detailed discussion and for methods to infer best matches in the presence of rate variations.

A best match (x, y) is reciprocal if (y, x) is also a best match. We will call a pair of reciprocal arcs (x, y) and (y, x) in a graph (G, σ) an *edge*, denoted by xy. In the absence of horizontal gene transfer, all pairs of orthologous genes form reciprocal best matches. That is, the undirected orthology graph is always a subgraph of the best match graph [7]. This simple observation has stimulated the search for computational methods to identify the "false-positive" edges in a BMG, i.e., edges that do not correspond to a pair of orthologous genes [16]. This requires a better understanding of the set of trees that explain a given BMG.

In this contribution, we derive an efficient algorithm for the construction of a binary tree that explains a BMG if and only if such a tree exists. Such BMGs will be called *binary-explainable*. This problem can be expressed as a consistency problem involving certain sets of both required and forbidden triples. It is therefore related to the MOST RESOLVED COMPATIBLE TREE and FORBIDDEN TRIPLES (RESTRICTED TO BINARY TREES) problems, both of which are NP-complete [2]. However, binary-explainable BMGs are characterized in [16] as those BMGs that do not contain a certain colored graph on four vertices, termed *hourglass*, as induced subgraph (Fig. 1E,F). The presence of an induced hourglass in an arbitrary vertex-colored graph $(G = (V, E), \sigma)$ can be checked in $O(|E|^2)$ [16]. This characterization, however, is not constructive and it remained an open problem how to construct a binary tree that explains a BMG.

This contribution is organized as follows. In Sect. 2, we introduce some notation and review key properties of BMGs that are needed later on. In Sect. 3, we derive a constructive algorithm for this problem that runs in near-cubic time $\tilde{O}(|V|^3)$. It produces a unique tree, the *binary-refinable tree* (BRT) of a BMG show in Fig. 3. The BRT has several interesting properties that are studied in detail in Sect. 4. Simulated data are used in Sect. 5 to show that BRTs are much better resolved than the least resolved trees of BMGs. Due to length restrictions, the proofs are omitted from this contribution. They can be found in `arxiv.org` preprint `2011.00511` [15].

2   Best Match Graphs

By construction, no vertex x of a BMG (G, σ) has a neighbor with the same color, i.e., the coloring σ is proper. Furthermore, every vertex x has at least one out-neighbor (i.e., a best match) y of every color $\sigma(y) \neq \sigma(x)$. We call such a proper coloring *sink-free* and say that (G, σ) is *sf-colored*.

We write $v \prec_T u$ for $v \preceq_T u$ and $v \neq u$ and use the convention that the vertices in an edge $uv \in E(T)$ are ordered such that $v \prec_T u$. Thus u is the unique *parent* of v, and v is a *child* of u. The set of all children of a vertex u is denoted by $\text{child}_T(u)$. The subtree of a tree T rooted at a vertex u is induced by the set of vertices $\{x \in V(T) \mid x \preceq_T u\}$ and will be denoted by $T(u)$.

A triple $ab|c$ is a rooted tree t on three pairwise distinct vertices $\{a, b, c\}$ such that $\text{lca}_t(a, b) \prec_t \text{lca}_t(a, c) = \text{lca}_t(b, c) = \rho$, where ρ denotes the root of t. A tree T' is *displayed* by a tree T, in symbols $T' \leq T$, if T' can be obtained from a subtree of T by contraction of edges [20]. Conversely, a tree T is a *refinement* of T' if $T' \leq T$ and additionally $L(T) = L(T')$. We denote by $r(T)$ the set of all triples that are displayed by a tree T, and write $\mathcal{R}_{|L} := \{xy|z \in \mathcal{R} : x, y, z \in L\}$ for the restriction of a triple set $\mathcal{R}$ to a set L of leaves.

A leaf-colored tree (T, σ) explaining a BMG (G, σ) is least resolved if every tree T' obtained from T by edge contractions no longer explains (G, σ). Thus, (T, σ) does not display a tree with fewer edges that explains (G, σ). As shown in [6], every BMG is explained by a unique least resolved tree.

Denoting by $G[L']$ the subgraph of G induced by a subset L' of vertices and by $\sigma_{|L'}$ the color map restricted to L', the following technical result relates induced subgraphs of BMGs to subtrees of their explaining trees:

Lemma 2 [14, Lemma 22]. *Let (G, σ) be a BMG explained by a tree (T, σ). Then, for every $u \in V(T)$, it holds $G(T(u), \sigma_{|L(T(u))}) = (G[L(T(u))], \sigma_{|L(T(u))})$.*

BMGs can be characterized in terms of informative and forbidden triples [6, 14, 17]. Given a vertex-colored graph (G, σ), we define

$$\begin{aligned}
\mathcal{R}(G, \sigma) &:= \{ab|b' : \sigma(a) \neq \sigma(b) = \sigma(b'), (a, b) \in E(G); (a, b') \notin E(G)\}, \\
\mathcal{F}(G, \sigma) &:= \{ab|b' : \sigma(a) \neq \sigma(b) = \sigma(b'), b \neq b'; (a, b), (a, b') \in E(G)\}.
\end{aligned} \tag{1}$$

We refer to $\mathcal{R}(G, \sigma)$ as the *informative triples* and to $\mathcal{F}(G, \sigma)$ as the *forbidden triples* of (G, σ). We will regularly make use of the observation that, as a direct consequence of their definition, forbidden triples always come in pairs:

Observation 1. *Let (G, σ) be a vertex-colored digraph. Then $ab|b' \in \mathcal{F}(G, \sigma)$ with $\sigma(b) = \sigma(b')$ if and only if $ab'|b \in \mathcal{F}(G, \sigma)$.*

Definition 3. *A pair of triple sets $(\mathcal{R}, \mathcal{F})$ is* consistent *if there is a tree T that displays all triples in $\mathcal{R}$ but none of the triples in $\mathcal{F}$. In this case, we say that T* agrees *with $(\mathcal{R}, \mathcal{F})$.*

For $\mathcal{F} = \emptyset$ this definition reduces to the usual notion of consistency of $\mathcal{R}$ [20]. In general, consistency of $(\mathcal{R}, \mathcal{F})$ can be checked in polynomial time. The algorithm MTT, named for *mixed triplets problem restricted to trees*, constructs a tree T that agrees with $(\mathcal{R}, \mathcal{F})$ or determines that no such tree exists [9]. It can be seen as a generalization of BUILD, which solves the corresponding problem for $\mathcal{F} = \emptyset$ [1]. Given a consistent triple set $\mathcal{R}$ on a set of leaves L, BUILD constructs a deterministic tree on L known as the *Aho tree*, and denoted here as $\text{Aho}(\mathcal{R}, L)$.

Two characterizations of BMGs given in [14, Theorem 15] and [17, Lemma 3.4 and Theorem 3.5] can be summarized as follows:

Proposition 4. *Let (G, σ) be a properly colored digraph with vertex set L. Then the following three statements are equivalent:*

1. *(G, σ) is a BMG.*
2. *$\mathcal{R}(G, \sigma)$ is consistent and $G(\mathrm{Aho}(\mathcal{R}(G, \sigma), L), \sigma) = (G, \sigma)$.*
3. *(G, σ) is sf-colored and $(\mathcal{R}(G, \sigma), \mathcal{F}(G, \sigma))$ is consistent.*

In this case, $(\mathrm{Aho}(\mathcal{R}(G, \sigma), L), \sigma)$ is the unique least resolved tree for (G, σ), and a leaf-colored tree (T, σ) on L explains (G, σ) if and only if it agrees with $(\mathcal{R}(G, \sigma), \mathcal{F}(G, \sigma))$.

3   Binary Trees Explaining a BMG in Near Cubic Time

We start with a few technical results on the structure of the triples sets $\mathcal{R}(G, \sigma)$ and $\mathcal{F}(G, \sigma)$.

Lemma 5. *Let (G, σ) be explained by a binary tree (T, σ). If $ab|b' \in \mathcal{F}(G, \sigma)$ with $\sigma(b) = \sigma(b')$, then (T, σ) displays the triple $bb'|a$.*

Lemma 5 implies that we can infer a set of additional triples that would be required for a binary tree to explain a vertex-colored graph (G, σ). This motivates the definition of an extended informative triple set

$$\mathcal{R}^B(G, \sigma) := \mathcal{R}(G, \sigma) \cup \{bb'|a \colon ab|b' \in \mathcal{F}(G, \sigma) \text{ and } \sigma(b) = \sigma(b')\}. \qquad (2)$$

Since informative and forbidden triples are defined by the presence and absence of certain arcs in a vertex-colored digraph, this leads to the following

Observation 2. *Let (G, σ) be a vertex-colored digraph and $L' \subseteq V(G)$. Then $R(G, \sigma)_{|L'} = R(G[L'], \sigma_{|L'})$ holds for any $R \in \{\mathcal{R}, \mathcal{F}, \mathcal{R}^B\}$.*

Lemma 6. *If (T, σ) is a binary tree explaining the BMG (G, σ), then (T, σ) displays $\mathcal{R}^B(G, \sigma)$.*

Lemma 7. *Let (G, σ) be an sf-colored digraph with vertex set L. Every tree on L that displays $\mathcal{R}^B(G, \sigma)$ explains (G, σ).*

Using Lemmas 6 and 7, it can be shown that consistency of $\mathcal{R}^B(G, \sigma)$ is sufficient for an sf-colored graph (G, σ) to be binary-explainable.

Theorem 8. *A properly vertex-colored graph (G, σ) with vertex set L is binary-explainable if and only if (i) (G, σ) is sf-colored, and (ii) $\mathcal{R}^B := \mathcal{R}^B(G, \sigma)$ is consistent. In this case, the BMG (G, σ) is explained by every refinement of the tree $(\mathrm{Aho}(\mathcal{R}^B, L), \sigma)$.*

Theorem 8 implies that the problem of determining whether an sf-colored graph (G, σ) is binary-explainable can be reduced to a triple consistency problem. More precisely, it establishes the correctness of Algorithm 1, which in turn relies on the construction of $\mathrm{Aho}(\mathcal{R}^B, L)$. The latter can be achieved in polynomial time [1]. Making use of the improvements achievable by using dynamic graph data structures [4, 10], we obtain the following performance bound:

Corollary 9. *There exists an $O(|L|^3 \log^2 |L|)$-time algorithm that constructs a binary tree explaining a vertex-colored digraph (G, σ) with vertex set L, if and only if such a tree exists.*

Algorithm 1: Construction of a binary tree explaining (G, σ).

Input: A properly vertex-colored graph (G, σ) with vertex set L.

Output: Binary tree (T, σ) explaining (G, σ) if one exists.

1 **if** (G, σ) *is not sf-colored* **then**

2 | **exit false**

3 construct the extended triple set $\mathcal{R}^{\mathrm{B}} := \mathcal{R}^{\mathrm{B}}(G, \sigma)$

4 $T \leftarrow \mathrm{Aho}(\mathcal{R}^{\mathrm{B}}, L)$

5 **if** T *is a tree* **then**

6 | construct an arbitrary binary refinement T' of T

7 | **return** (T', σ)

8 **else**

9 | **exit false**

---

4 The Binary-Refinable Tree of a BMG

If a graph (G, σ) with vertex set L is binary-explainable, Theorem 8 implies that $\mathcal{R}^{\mathrm{B}} := \mathcal{R}^{\mathrm{B}}(G, \sigma)$ is consistent and every refinement of $(\mathrm{Aho}(\mathcal{R}^{\mathrm{B}}, L), \sigma)$ explains (G, σ). In this section, we investigate the properties of this tree in more detail.

Definition 10. *The* binary-refinable tree (BRT) *of a binary-explainable BMG* (G, σ) *with vertex set* L *is the leaf-colored tree* $(\mathrm{Aho}(\mathcal{R}^{B}(G, \sigma), L), \sigma)$.

The BRT is not necessarily a binary tree. However, Theorem 8 implies that the BRT as well as each of its binary refinements explains (G, σ). Note that the tree $\mathrm{Aho}(\mathcal{R}^{\mathrm{B}}, L)$ and thus the BRT are well-defined because Theorem 8 ensures consistency of $\mathcal{R}^{\mathrm{B}}$ for binary-explainable graphs, and the Aho tree as produced by BUILD is uniquely determined by the set of input triples [1].

Corollary 11. *If* (G, σ) *is a binary explainable BMG, then its BRT is a refinement of the LRT.*

Clearly, the BRT is least resolved among the trees that display $\mathcal{R}^{\mathrm{B}}$, i.e., contraction of an arbitrary edge results in a tree that no longer displays all triples in $\mathcal{R}^{\mathrm{B}}$ [19, Proposition 4.1]. Now, we tackle the question whether the BRT is the unique least resolved tree in this sense. In other words, we ask whether every tree that displays $\mathcal{R}^{\mathrm{B}}$ is a refinement of the BRT. As we shall see, this question can be answered in the affirmative.

In order to show this, we first introduce some additional notation and concepts for sets of triples. Following [3,18], we call the *span* of $\mathcal{R}$, denoted by $\langle \mathcal{R} \rangle$, the set of all trees with leaf set $L_{\mathcal{R}} := \bigcup_{t \in \mathcal{R}} L(t)$ that display $\mathcal{R}$. With this notion, we define the *closure operator* for consistent triple sets by

$$\mathrm{cl}(\mathcal{R}) = \bigcap_{T \in \langle \mathcal{R} \rangle} r(T), \tag{3}$$

i.e., a triple t is contained in $\mathrm{cl}(\mathcal{R})$ if all trees that display $\mathcal{R}$ also display t. In particular, $\mathrm{cl}(\mathcal{R})$ is again consistent. The map cl is a closure in the usual sense on the set of consistent triple sets, i.e., it is extensive [$\mathcal{R} \subseteq \mathrm{cl}(\mathcal{R})$], monotonic [$\mathcal{R}' \subseteq \mathcal{R} \implies \mathrm{cl}(\mathcal{R}') \subseteq \mathrm{cl}(\mathcal{R})$], and idempotent [$\mathrm{cl}(\mathcal{R}) = \mathrm{cl}(\mathrm{cl}(\mathcal{R}))$] [3, Proposition 4]. A consistent set of triples $\mathcal{R}$ is *closed* if $\mathcal{R} = \mathrm{cl}(\mathcal{R})$.

Interesting properties of a triple set $\mathcal{R}$ and of the Aho tree $\mathrm{Aho}(\mathcal{R}, L)$ can be understood by considering the *Aho graph* $[\mathcal{R}, L]$ with vertex set L and edges xy iff there is a triple $xy|z \in \mathcal{R}$ with $x, y, z \in L$ [1]. It has been shown in [3] that a triple set $\mathcal{R}$ on L is consistent if and only if $[\mathcal{R}_{|L'}, L']$ is disconnected for every subset $L' \subseteq L$ with $|L'| > 1$. The root ρ of the Aho tree $\mathrm{Aho}(\mathcal{R}, L)$ corresponds to the Aho graph $[\mathcal{R}, L]$ in such a way that there is a one-to-one correspondence between the children v of ρ and the connected components C of $[\mathcal{R}, L]$ given by $L(T(v)) = V(C)$. The BUILD algorithm constructs $\mathrm{Aho}(\mathcal{R}, L)$ by recursing top-down over the connected components (with vertex sets L') of the Aho graphs. It fails if and only if $|L'| > 1$ and $[\mathcal{R}_{|L'}, L']$ is connected at some recursion step. For a more detailed description we refer to [1]. Since the decomposition of the Aho graphs into their connected components is unique, the Aho tree is also uniquely defined.

The following characterization of triples that are contained in the closure also relies on Aho graphs:

Proposition 12 [2, Corollary 3.9]. *Let $\mathcal{R}$ be a consistent set of triples and $L_{\mathcal{R}} := \bigcup_{t \in \mathcal{R}} L(t)$ the union of their leaves. Then $ab|c \in \mathrm{cl}(\mathcal{R})$ if and only if there is a subset $L' \subseteq L_{\mathcal{R}}$ such that the Aho graph $[\mathcal{R}_{|L'}, L']$ has exactly two connected components, one containing both a and b, and the other containing c.*

The following result shows that Proposition 12 can be applied to the triple set $\mathcal{R}^B(G, \sigma)$ of an sf-colored graph (G, σ) with the exception of the two trivial special cases in which either all vertices of (G, σ) are of the same color or of pairwise distinct colors.

Lemma 13. *Let (G, σ) be an sf-colored graph with vertex set $L \neq \emptyset$, $L_{\mathcal{R}, \mathcal{F}} := \bigcup_{t \in \mathcal{R}(G, \sigma) \cup \mathcal{F}(G, \sigma)} L(t)$ and $L_{\mathcal{R}^B} := \bigcup_{t \in \mathcal{R}^B(G, \sigma)} L(t)$. Then the following statements are equivalent:*

1. *$L_{\mathcal{R}, \mathcal{F}} = L_{\mathcal{R}^B} = L$.*
2. *$\mathcal{R}(G, \sigma) \cup \mathcal{F}(G, \sigma) \neq \emptyset$.*
3. *$\mathcal{R}^B(G, \sigma) \neq \emptyset$.*
4. *(G, σ) is ℓ-colored with $\ell \geq 2$ and contains two vertices of the same color.*

If these statements are not satisfied, then (G, σ) is a BMG that is explained by any tree (T, σ) on L.

Lemma 13 holds for BMGs since these are sf-colored by Proposition 4. The following result is essential for the application of Proposition 12 to a triple set $\mathcal{R}^B(G, \sigma)$.

Lemma 14. *Let (G, σ) be a binary-explainable BMG with vertex set L and $\mathcal{R}^B := \mathcal{R}^B(G, \sigma)$. Then, for any two distinct connected components C and C' of the Aho graph $H := [\mathcal{R}^B, L]$, the subgraph $H[L']$ induced by $L' = V(C) \uplus V(C')$ satisfies $H[L'] = [\mathcal{R}^B_{|L'}, L'] = C \uplus C'$.*

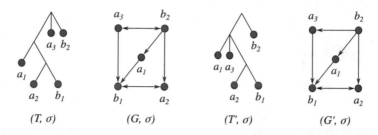

Fig. 2. A least resolved tree (T, σ) explaining the BMG (G, σ) with informative triples $\mathcal{R} := \mathcal{R}(G, \sigma) = \{a_2b_1|a_1, a_2b_1|a_3, a_2b_1|b_2, a_1b_1|b_2\}$ for which $r(T) \neq \mathrm{cl}(\mathcal{R})$. The tree (T', σ) also displays $\mathcal{R}$ but $a_1a_2|a_3 \in r(T)$ and $a_1a_2|a_3 \notin r(T')$. In particular, (T', σ) explains a different BMG (G', σ) in which the arc (a_3, b_2) is missing.

Lemma 15. *The BRT (T, σ) of a binary-explainable BMG (G, σ) satisfies $r(T) = \mathrm{cl}(\mathcal{R}^B(G, \sigma))$.*

No analog of Lemma 15 holds for LRTs, i.e., in general we have $\mathrm{cl}(\mathcal{R}(G, \sigma)) \neq r(T)$ for the LRT (T, σ) of a BMG (G, σ). Figure 2 shows a counterexample.

Following [5,8], a set of rooted triples $\mathcal{R}$ *identifies* a tree T on L if T displays $\mathcal{R}$ and every other tree on L that displays $\mathcal{R}$ is a refinement of T.

Proposition 16 [8, Lemma 2.1]. *Let T be a phylogenetic tree and $\mathcal{R} \subseteq r(T)$. Then $\mathrm{cl}(\mathcal{R}) = r(T)$ if and only if $\mathcal{R}$ identifies T.*

From Lemma 15 and Proposition 16 we immediately obtain the main result of this section:

Theorem 17. *Let (G, σ) be a binary-explainable BMG with vertex set L and BRT (T, σ). Then every tree on L that displays $\mathcal{R}^B(G, \sigma)$ is a refinement of (T, σ). In particular, every binary tree that explains (G, σ) is a refinement of (T, σ).*

Corollary 18. *If (G, σ) is binary-explainable with BRT (T, σ), then a binary tree (T', σ) explains (G, σ) if and only if it is a refinement of (T, σ).*

Assuming that evolution of a gene family only progresses by bifurcations and that the correct BMG (G, σ) is known, Corollary 18 implies that the true (binary) gene tree displays the BRT of (G, σ). Figure 3 shows the LRT and BRT for the BMG (G, σ) in Fig. 1C. The BRT is more finely resolved than the LRT, see also Fig. 1D. The difference arises from the triple $a_2a_3|c_2 \in \mathcal{R}^B(G, \sigma) \setminus \mathcal{R}(G, \sigma)$. The true gene tree in Fig. 1(A,B) is a binary refinement of the BRT (and thus also of the LRT).

5   Simulation Results

Best match graphs contain valuable information on the (rooted) gene tree topology since both their LRTs and BRTs are displayed by the latter (cf. [6] and

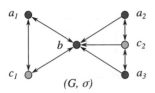 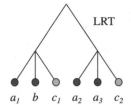 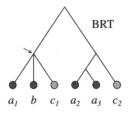

Fig. 3. The binary-refinable tree (BRT) of the binary-explainable BMG (G, σ) (cf. Fig. 1C) is better resolved than its LRT (cf. Fig. 1C). The remaining polytomy in the BRT (red arrow) can be resolved arbitrarily. Out of the three possibilities, one results in the original binary tree (cf. Fig. 1B).

Corollary 18). Hence, they are of interest for the reconstruction of gene family histories. In order to illustrate the potential benefit of using the better resolved BRT instead of the LRT, we simulated realistic, but idealized, evolutionary scenarios using the library `AsymmeTree` [23], i.e., we extracted the "true" BMGs from the simulated gene trees. Hence, we do not take into account errors arising in the approximation of best matches from sequence data. In real-life applications, of course, factors such as rate variation among different branches and inaccuracies in sequence alignment need to be taken into account, see e.g. [17,23] for a more detailed discussion of this topic.

In brief, species trees are generated using the Innovation Model [11]. A so-called planted edge above the root is added to account for the ancestral line, in which gene duplications may already occur. This planted tree S is then equipped with a dating function that assigns time stamps to its vertices. Binary gene trees $\widetilde{T}$ are simulated along the edges of the species tree by means of a constant-rate birth-death process extended by additional branchings at the speciations. For HGT events, the recipient branches are assigned at random. An extant gene x corresponds to a branch of $\widetilde{T}$ that reaches present time and thus a leaf s of S, determining $\sigma(x) = s$. All other leaves of $\widetilde{T}$ correspond to losses. To avoid trivial cases, losses are constrained in such a way that every branch (and in particular every leaf) of S has at least one surviving gene. The observable part T of $\widetilde{T}$ is obtained by removing all branches that lead to losses only and suppressing inner vertices with a single child. From (T, σ), the BMG and its LRT and BRT are constructed.

We consider single leaves and the full set L as trivial clades since they appear in any phylogenetic tree $T = (V, E)$ with leaf set L. We can quantify the resolution $\text{res}(T)$ as the fraction of non-trivial clades of T retained in the LRT or BRT, respectively, which is the same as the fraction of inner edges that remain uncontracted. To see this, we note that T has between 0 and $|L| - 2$ edges that are not incident with leaves, with the maximum attained if and only if T is binary. Thus T has $|E| - |L|$ edges that have remained uncontracted. On the other hand, each vertex of T that is not a leaf or the root defines a non-trivial clade. Thus T contains $|V| - 1 - |L|$ non-trivial clades. Since $|E| = |V| - 1$ we

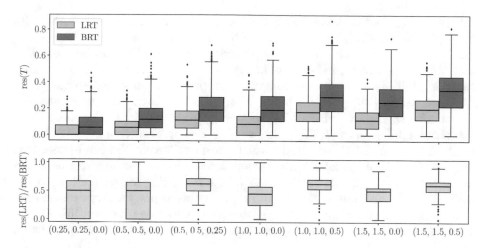

Fig. 4. Comparison of LRTs and BRTs of BMGs obtained from simulated evolutionary scenarios with 10 to 30 species and binary gene trees with different combinations of rates for gene duplications, gene loss, and horizontal transfer (indicated as triples on the horizontal axis). Top: Fraction of resolved non-trivial clades res(LRT) and res(LRT). Below: The ratio of these parameters. Distributions are computed from 1000 scenarios for each combination of rates. The box plots show the median and inter-quartile range.

have

$$\text{res}(T) := \frac{|E| - |L|}{|L| - 2} = \frac{|V| - |L| - 1}{|L| - 2}. \tag{4}$$

The parameter res(T) is well-defined for $|L| > 2$, which is always the case in the simulated scenarios. It satisfies res$(T) = 0$ for a tree consisting only of the root and leaves, and res$(T) = 1$ for binary trees. Since the true gene tree (T, σ) is binary, it displays both the LRT and BRT of its BMG. Thus we have $0 \leq \text{res}(\text{LRT}) \leq \text{res}(\text{BRT}) \leq \text{res}(T) = 1$.

The results for the simulated scenarios with different rates for duplications, losses, and horizontal transfers are summarized in Fig. 4. In general, the BRT is much better resolved than the LRT with the median values of res(BRT) exceeding res(LRT) by about a factor of two (cf. lower panel).

6   Concluding Remarks

We have shown here that binary-explainable BMGs are explained by a unique binary-resolvable tree (BRT), which displays the also unique least resolved tree (LRT). In general, the BRT differs from the LRT. All binary explanations are obtained by resolving the multifurcations in the BRT in an arbitrary manner. The constructive characterization of binary-explainable BMGs given here can be computed in near-cubic time, improving the quartic-time non-constructive characterization in [16], which is based on the hourglass being a forbidden induced subgraph. We note that binarizing a leaf-colored tree (T, σ) does not affect its "biological feasibility", i.e., the existence of a reconciliation map $\mu : T \to S$ to a given or unknown species tree S, since every gene tree (T, σ) can be reconciled with any species tree on the set $\sigma(L(T))$, see e.g. [7]. We can therefore safely use the additional information contained in the BRT compared to the LRT. As discussed in [16], poor resolution of the LRT is often the consequence of consecutive speciations without intervening gene duplications. The same argumentation applies to the BRT, which we have seen in Sect. 5 to be much better resolved than the LRT. Still BRTs usually are not binary. We can expect that the combination of the BRT with *a priori* knowledge on the species tree S can be used to unambiguously resolve most of the remaining multifurcations in the BRT. The efficient computation of BRTs is therefore of potentially practical relevance whenever evolutionary scenarios are essentially free from multifurcations, an assumption that is commonly made in phylogenetics but may not always reflect the reality [22].

Acknowledgments. This work was supported in part by the Austrian Federal Ministries BMK and BMDW and the Province of Upper Austria in the frame of the COMET Programme managed by FFG.

References

1. Aho, A.V., Sagiv, Y., Szymanski, T.G., Ullman, J.D.: Inferring a tree from lowest common ancestors with an application to the optimization of relational expressions. SIAM J. Comput. **10**, 405–421 (1981). https://doi.org/10.1137/0210030
2. Bryant, D.: Building Trees, Hunting for Trees, and Comparing Trees: Theory and Methods in Phylogenetic Analysis. Dissertation, University of Canterbury, Canterbury, NZ (1997)
3. Bryant, D., Steel, M.: Extension operations on sets of leaf-labeled trees. Adv. Appl. Math. **16**(4), 425–453 (1995). https://doi.org/10.1006/aama.1995.1020
4. Deng, Y., Fernández-Baca, D.: Fast compatibility testing for rooted phylogenetic trees. Algorithmica **80**(8), 2453–2477 (2017). https://doi.org/10.1007/s00453-017-0330-4
5. Geiß, M., Anders, J., Stadler, P.F., Wieseke, N., Hellmuth, M.: Reconstructing gene trees from Fitch's xenology relation. J. Math. Biol. **77**(5), 1459–1491 (2018). https://doi.org/10.1007/s00285-018-1260-8
6. Geiß, M., et al.: Best match graphs. J. Math. Biol. **78**(7), 2015–2057 (2019). https://doi.org/10.1007/s00285-019-01332-9
7. Geiß, M., et al.: Best match graphs and reconciliation of gene trees with species trees. J. Math. Biol. **80**(5), 1459–1495 (2020). https://doi.org/10.1007/s00285-020-01469-y
8. Grünewald, S., Steel, M., Swenson, M.S.: Closure operations in phylogenetics. Math. Biosci. **208**, 521–537 (2007). https://doi.org/10.1016/j.mbs.2006.11.005
9. He, Y.J., Huynh, T.N.D., Jansson, J., Sung, W.K.: Inferring phylogenetic relationships avoiding forbidden rooted triplets. J. Bioinf. Comp. Biol. **4**, 59–74 (2006). https://doi.org/10.1142/S0219720006001709
10. Henzinger, M.R., King, V., Warnow, T.: Constructing a tree from homeomorphic subtrees, with applications to computational evolutionary biology. Algorithmica **24**, 1–13 (1999). https://doi.org/10.1007/PL00009268

11. Keller-Schmidt, S., Klemm, K.: A model of macroevolution as a branching process based on innovations. Adv. Complex Syst. **15**, 1250043 (2012). https://doi.org/10.1142/S0219525912500439

12. Nichio, B.T.L., Marchaukoski, J.N., Raittz, R.T.: New tools in orthology analysis: a brief review of promising perspectives. Front. Genet. **8**, 165 (2017). https://doi.org/10.3389/fgene.2017.00165

13. Rusin, L.Y., Lyubetskaya, E., Gorbunov, K.Y., Lyubetsky, V.: Reconciliation of gene and species trees. BioMed. Res. Int. **2014** (2014). https://doi.org/10.1155/2014/642089

14. Schaller, D., et al.: Corrigendum to "Best Match Graphs". J. Math. Biol. In press (2021). arxiv.org/1803.10989v4, https://doi.org/10.1016/j.tcs.2021.02.037

15. Schaller, D., Geiß, M., Hellmuth, M., Stadler, P.F.: Best Match Graphs with Binary Trees. Tech. Rep. 2011.00511 arxiv.org/2011.00511 (2020)

16. Schaller, D., Geiß, M., Stadler, P.F., Hellmuth, M.: Complete characterization of incorrect orthology assignments in best match graphs. J. Math. Biol. **82**(3), 1–64 (2021). https://doi.org/10.1007/s00285-021-01564-8

17. Schaller, D., Stadler, P.F., Hellmuth, M.: Complexity of modification problems for best match graphs. Theor. Comp. Sci. **865**, 63–84 (2021). https://doi.org/10.1016/j.tcs.2021.02.037

18. Seemann, C.R., Hellmuth, M.: The matroid structure of representative triple sets and triple-closure computation. Eur. J. Comb. **70**, 384–407 (2018). https://doi.org/10.1016/j.ejc.2018.02.013

19. Semple, C.: Reconstructing minimal rooted trees. Discr. Appl. Math. **127**, 489–503 (2003). https://doi.org/10.1016/S0166-218X(02)00250-0

20. Semple, C., Steel, M.: Phylogenetics. Oxford University Press, Oxford UK (2003)

21. Setubal, J.C., Stadler, P.F.: Gene phylogenies and orthologous groups. In: Setubal, J.C., Stadler, P.F., Stoye, J. (eds.) Comparative Genomics, vol. 1704, pp. 1–28. Springer, Heidelberg (2018). https://doi.org/10.1007/978-1-4939-7463-4_1

22. Slowinski, J.B.: Molecular polytomies. Mol. Phylogenet. Evol. **19**(1), 114–120 (2001). https://doi.org/10.1006/mpev.2000.0897

23. Stadler, P.F., et al.: From pairs of most similar sequences to phylogenetic best matches. Alg. Mol. Biol. **15**, 5 (2020). https://doi.org/10.1186/s13015-020-00165-2

Scalable and Accurate Phylogenetic Placement Using pplacer-XR

Eleanor Wedell$^{(\boxtimes)}$ ⓘ, Yirong Cai ⓘ, and Tandy Warnow$^{(\boxtimes)}$ ⓘ

Department of Computer Science, University of Illinois at Urbana-Champaign,
Urbana, IL, USA
{ewedell2,yirong2,warnow}@illinois.edu

Abstract. Phylogenetic placement, the problem of placing a sequence into a precomputed phylogenetic "backbone" tree, is useful for constructing large trees, performing taxon identification of newly obtained sequences, and other applications. The most accurate current method, pplacer, performs the placement using maximum likelihood but fails frequently on backbone trees with 5000 sequences. We show a simple technique, pplacer-XR (pplacer-eXtra Range), that extends pplacer to large datasets. We show, using challenging large datasets, that pplacer-XR provides the accuracy of pplacer and the scalability to ultra-large datasets of a leading fast phylogenetic placmement method, APPLES. pplacer-XR is available in open source form on github.

Keywords: Phylogenetic placement · Likelihood-based methods · Phylogenetics

1 Introduction

Phylogenetic placement is the process of taking a sequence (called a "query sequence") and adding it into a phylogenetic tree (called the "backbone tree"). These methods are used for taxonomic identification, obtaining microbiome profiles, and biodiversity assessment [5–7,9,12,14]. Furthermore, phylogenetic placement can be used to update very large phylogenies [2], where they offer a computationally feasible approach in comparison to *de novo* phylogeny estimation (which is NP-hard in most formulations).

One of the earliest methods for phylogenetic placement is pplacer [8], which uses a maximum likelihood approach (as well as a Bayesian method) to add query sequences into the tree. The input is a query sequence and a backbone tree, as well as a multiple sequence alignment of the sequences at the leaves of the backbone tree as well as the query sequence. The backbone tree is also given with numeric parameters (branch lengths, substitution rate matrix, etc.) for the selected sequence evolution model (e.g., the Generalized Time Reversible model [16]). Given this input, pplacer finds the best edge (under the maximum

---

Y. Cai and E. Wedell—Contributed equally to this work.

© Springer Nature Switzerland AG 2021
C. Martín-Vide et al. (Eds.): AlCoB 2021, LNBI 12715, pp. 94–105, 2021.
https://doi.org/10.1007/978-3-030-74432-8_7

likelihood criterion) in the backbone tree to attach the query sequence to, then subdivides the edge and returns the enlarged tree as well as the updated branch lengths, and a confidence score for the placements returned. EPA [4] was developed in 2011 with the same likelihood-based approach and accuracy, and was replaced in 2019 by EPA-ng [3], an improved version of EPA. In 2020, Balaban et al. [2] developed APPLES, a distance-based approach to phylogenetic placement. Balaban et al. evaluated EPA-ng, pplacer, and APPLES on a range of model conditions and revealed that while EPA-ng and pplacer provide high accuracy (with pplacer more accurate than EPA-ng), only APPLES was able to run on large backbone trees. Furthermore, pplacer, although the most accurate of the three methods, was limited to backbone trees with only a few thousand sequences, and often failed in [2] when the backbone tree had 5000 sequences. When pplacer failed in [2], these occurred when pplacer was able to finish running but produced placements with confidence scores of $-\infty$ (negative infinity). We are also aware that pplacer has high memory requirements, as reported in the github site for GTDB-tk [6]. Thus, pplacer, currently the most accurate placement method available, may be limited to small backbone trees (unless, perhaps, there is access to large amounts of memory), while APPLES, the most scalable method, has reduced accuracy compared to both EPA-ng and pplacer.

To scale pplacer to larger datasets, we use the following strategy, which we call "pplacer-XR" (pplacer-eXtra Range). Rather than attempting to find the best location in the entire backbone tree into which we insert the query sequence, pplacer-XR uses an informed strategy to select a subtree of the backbone tree, places the query sequence into that subtree, and then identifies the correct location in the backbone tree associated with that location. The first and third of these steps involve using distances, and the middle step uses pplacer; hence pplacer-XR is an extension of pplacer to enable it to run on larger backbone trees (and is identical to pplacer on those backbone trees that are small enough for pplacer to run). This approach is unique amongst placement methods, in that it focuses on placement locally among a subset of taxa, rather than the performing statistical analysis on the entire set of sequences. Our experimental study, using the datasets from the study presenting APPLES [2], shows that pplacer-XR, although slower than APPLES, improves on the accuracy of APPLES and matches its scalability. Our study also shows that pplacer-XR is more accurate and also faster than EPA-ng (which, like pplacer, does not scale to large backbone trees). Thus, pplacer-XR is the most accurate of the existing phylogenetic placement methods, and matches the scalability (but not the speed) of the most scalable of these methods.

The remainder of the paper is organized as follows. In Sect. 2 we present the pplacer-XR method. We describe the experimental study in Sect. 3. The results of the study are presented in Sect. 4, and we close with a discussion in Sect. 5.

2   pplacer-XR

The input to pplacer-XR is (1) a backbone tree T with a set S of sequences labelling the leaves, (2) a query sequence q, (3) a multiple sequence alignment A

(of length k) that includes all of the sequences in S as well as the query sequence q, and (4) a parameter B (to be the largest size that pplacer can reliably run on). We describe pplacer-XR for a single query sequence, as placing a set of sequences can be done independently. We also assume that the backbone tree is provided with the numeric parameters (branch lengths, substitution rate matrix, etc., as computed using RAxML [15]), associated with the specified sequence evolution model. However, if the backbone tree does not have these numeric parameters, we use RAxML to estimate the parameters (and will need to resolve the tree if it is not binary before using RAxML). We assume that pplacer-XR is used to add nucleotide sequences into a backbone tree under the GTR [16] model, but we note that other models could be used without any important change to the method. At a high level, our three-stage technique operates as follows:

- Stage 1: A subtree T' of T is identified (defined by its set S' of leaves), with the restriction that T' cannot contain more than B sequences (and we set $B = 2000$ by default). This is referred to as the "placement tree".
- Stage 2: We use pplacer to find the best edge e' of T' for the query sequence.
- Stage 3: Since e' may correspond to a path with more than one edge in T, we use distances (estimated by pplacer) to find the correct edge of T into which we add the query sequence.

Stage 1: A key issue in Stage 1 is how to set B, which limits the size of the placement tree. In general, larger placement trees provide better accuracy (as shown in [11]), but also run the risk of pplacer failing (due to the backbone tree being too large) and will increase the running time. Hence, we have set $B = 2000$, which is a size that pplacer reliably completes on. Note that if the backbone tree is small enough (i.e., has at most B leaves), then pplacer-XR defaults to pplacer; hence this algorithm only applies when the backbone tree has more than B leaves.

In order to obtain a subtree most likely to contain the correct placement, pplacer-XR uses a greedy strategy to find a set of B leaves in T that are close to the query sequence. The first step is to find a closest leaf l (using the Hamming distance to the query sequence, which is well defined since we are given a multiple sequence alignment). Hence, this step takes $O(nk)$ time.

Once that leaf l is found, pplacer-XR uses a breadth first search to find $B-1$ additional leaves, here based on the path length to the leaf l, defined as follows. The path distance in T from a given leaf l' to l is $\sum_{e_i \in P} L(e_i)$ where P is the path in T from l to l' and $L(e_i)$ is the length of the edge e_i in P. Starting from l, we use a breadth-first search to select those leaves in T that have the lowest path distance to l until we reach B leaves, and we return the subtree of T induced by this set of leaves. Once the set of B leaves is identified, the induced subtree T' is returned, with the branch lengths in T' computed by using the associated branch lengths in T. Hence, Stage 1 takes $O(nk)$ time, and returns a set of B leaves and the induced subtree T' (with its associated branch lengths and other numeric parameters), which will be the placement tree passed to pplacer in Stage 2.

Stage 2: We then run pplacer on the placement tree T' we obtain from Stage 1. This identifies an edge e' in T', which will correspond to either a single edge e or a path $P(e')$ of two or more edges in T. Since a single edge is vacuously a path, we will let $P(e')$ denote the edge or path in T corresponding to e'. To determine $P(e')$ given e', note that e' defines a bipartition $\pi(e')$ on T'. At least one, and possibly more than one, of the edges in T define bipartitions that are compatible with $\pi(e)$ (meaning specifically that they induce the same bipartition when restricted to the leafset of T'). The set of edges in T that define bipartitions compatible with $\pi(e')$ form either a single edge or a path, and defines $P(e')$. We then set $L((e')$ (i.e., the length of edge e') to be $L(P(e'))$, where $L(P(e'))$ is the sum of the branch lengths in the path (or edge) in T denoted by $P(e')$.

Stage 3: If e' corresponds to a single edge e in T, then we place the query sequence into that edge. However, if e' corresponds to a path with two or more edges in T, then we use the distances we obtained to find the correct placement edge for the query sequence, as we now describe.

Recall that the tree T' is a subtree of T formed by specifying a set of leaves, and that the edges of T' have branch lengths that correspond to the branch lengths in T. Specifically, if an edge in T' also exists in T, then they have the same branch length, and if the edge e' in T' corresponds to a path P in T, then the length of the edge e' is the sum of the lengths of the edges in P. Recall also that when pplacer inserts the query sequence into e' in T', it also subdivides the edge e' and specifies how the branch length is divided. For example, suppose $e' = (a, b)$ is an edge in T' with length $L(e')$, and the query sequence is attached to this edge. Then pplacer subdivides the edge e', thus creating two new edges (a, v) and (v, b), whose lengths add up to $L(e')$. We then use those new lengths to determine exactly what edge in T we should insert the query sequence into, and how to divide the length of that edge to produce the desired outcome.

3   Experimental Study Design

Overview. We used simulated datasets from [2] to explore pplacer-XR in comparison to other phylogenetic placement methods. In these analyses, the available sequences were evolved down a model tree and backbone trees were constructed on these sequences. Thus, query sequences (also generated in the simulation) can be added into the backbone trees using phylogenetic placement methods. The true tree on every subset of the sequences is therefore known, and so error produced by a phylogenetic placement method can be exactly quantified. We used the leave-one-out approach from [2] to evaluate accuracy. Specifically, Balaban et al. [2] provided 200 query sequences for each model condition they studied. Given query sequence q, we extract the leaf for q from the initial backbone tree, thus producing the reduced backbone tree which we denote by T. We then use the placement method to insert q into T, thus producing a tree with the same set of leaves as the initial backbone tree.

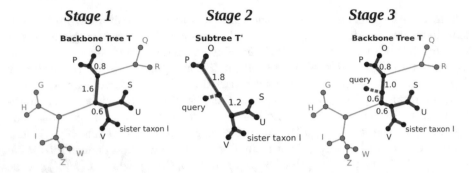

Fig. 1. Description of the pplacer-XR technique. In Stage 1, we select the placement subtree T' from the backbone tree T, for a specified query sequence. To find the placement subtree T' of T, we first find the leaf l with the smallest Hamming distance to the query sequence (here called the "sister taxon"). Then, we greedily pick the $B-1$ leaves (here $B = 6$) that have the smallest path distance to l (using branch lengths to define path distances). In this case, we select five leaves O,P,S,U,V, and the placement subtree T' is induced by the set $\{P, O, S, U, V, l\}$ of six leaves. Here we show pplacer selecting an edge in T' separating leaves $\{P, O\}$ from $\{S, U, V, l\}$, and this single edge in T' corresponds to a path of three edges in T. Note that pplacer returns not only which edge in the placement subtree to insert the query sequence into, but the branch lengths on either side; this is used to find the correct placement of the query sequence in Stage 3.

Because pplacer has already been established in [2] to be the most accurate method when it can run and because pplacer and pplacer-XR are the same when pplacer can run, we only explore results on datasets that are too big for pplacer. Therefore, we selected the larger backbone trees from the study by Balaban et al. [2] to evaluate pplacer-XR. Specifically, we examine a set of "variable size" backbone trees with 5000 to 200,000 leaves, each a subset of the RNASim [10] million-sequence simulated dataset (which evolved under a complex biophysical model that includes selection). The backbone trees for these datasets were estimated using FastTree2 on the RNASim alignments.

We compare pplacer-XR to EPA-ng v0.3.8 using the default settings under the GTR model [16] with the Γ model of rate variation [17], and APPLES v1.3.0 run using its default settings. We ran all analyses on the University of Illinois Campus Cluster, which limits all analyses to four hours. Therefore, any analyses that did not complete within four hours were marked as "failures".

Backbone Trees and Branch Lengths. We use the RNASim variable size datasets studied in Balaban et al. [2], with the same backbone trees (topologies and branch lengths), alignments, and query sequences. These variable size datasets have backbone trees computed on the true alignment using FastTree2 [13] and contain 5,000, 10,000, 50,000 and 100,000 leaves (each with five replicates) and 200,000 (one replicate). We use the query sequences provided in [2] for each backbone tree; each such tree has 200 query sequences, which are drawn from the backbone tree, and thus are in the original RNAsim dataset. The true alignment for the RNASim dataset, modified to remove the sites with more than 95% gaps, is provided to all phylogenetic placement methods. Balaban et al. computed branch lengths on all the backbone trees using protocols specific to each placement method (e.g., FastME for APPLES and RAxML for EPA-ng and pplacer); we use their backbone trees and branch lengths for EPA-ng and APPLES, and use the same command for RAxML to compute branch lengths for pplacer-XR as they used for pplacer.

Performance Criteria. We report placement error using the "delta error", as used by Balaban et al. [2] (described below). We also report the running time, recognizing that since the methods were run on a heterogeneous system (the University of Illinois Campus Cluster), these are not exactly comparable.

The delta error is defined using the concept of "false negatives", which we now define. Note that every edge in an unrooted tree t defines a bipartition on its leafset, and the set $B(t)$ of these bipartitions defines the tree t. Now let T^* denote the true tree on n leaves and let T denote a tree estimated on a subset $\mathcal{L}$ of $n - 1$ of the leaves of T^*. We can restrict T^* to the leaves $\mathcal{L}$ of T, and denote this tree by $T^*|_{\mathcal{L}}$. Then, the *false negatives* of the estimated tree T with respect to the true tree T^* are the bipartitions in $B(T^*|_{\mathcal{L}}) \setminus B(T)$. When we insert a query sequence into tree T we obtain a tree P with one additional leaf (and hence the same number of leaves as T^*); the number of false negatives can therefore go up or stay the same, but cannot decrease. This means that the delta error is always non-negative. It is, however, possible for the delta error to be 0 without any of the estimated trees being correct (e.g., imagine the case where the backbone tree is not identical to the true tree, but the query sequence is correctly placed, perhaps as the sibling to another leaf; in such a case, the delta error will be 0 although the backbone tree before and after placement has errors). Put formally, the delta error produced by adding the missing query sequence into the backbone tree T (which is lacking one leaf) to obtain tree P is

$$\Delta e(P) = |B(T^*) \setminus B(P)| - |B(T^*|_{\mathcal{L}}) \setminus B(T)|. \tag{1}$$

We report the average and maximum delta error as well as the number of placements with a delta error of 0.

4   Results

All three methods were run on all datasets. APPLES and pplacer-XR successfully completed on all datasets, including the analyses with backbone trees containing 200,000 leaves. However, results with EPA-ng on backbone trees of size 50,000 were not obtained for just over 800 placements due to a core dump while running. This seemed to be a result of memory limitations, rather than time limitations. As a result, for the backbone trees larger than 10,000 taxa pplacer-XR is compared only to APPLES. All other placement methods (except EPA-ng) were able to complete within the 4-hr time limit.

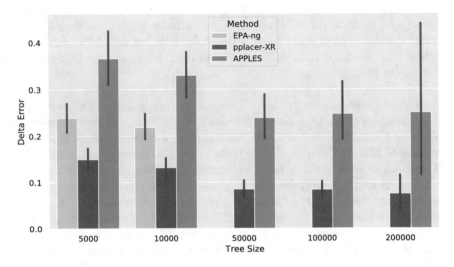

Fig. 2. Delta error (means plus standard error) over of all placements for each method over five backbone tree sizes. Averages for the 5,000, 10,000, 50,000, and 100,000 taxa trees are over all 1,000 query sequence placements, and for the 200,000 taxa tree the averages are for 200 query sequence placements. EPA-ng was unable to run for trees larger than 10,000 taxa and so no results are reported for EPA-ng on those datasets. This was run on the UIUC Campus Cluster with 64 GB of memory per placement and 4 h time limit.

4.1   Query Placement Accuracy

Figure 2 and Table 2 show that the mean delta error was significantly lower for the pplacer-XR than for APPLES and EPA-ng. EPA-ng was more accurate than APPLES on those datasets on which it completed, but could only reliably complete on the 5000- and 10,000-sequence backbone trees. The delta error results indicate that for all methods the accuracy improves as the backbone tree size increases, suggesting that denser taxon sampling improves phylogenetic placement accuracy. However, for every backbone tree size, APPLES has 2–3 times the mean delta error of pplacer-XR.

Figure 3 and Table 3 show the maximum delta error in each dataset, which corresponds to the placement with the highest delta error in that set. Note that pplacer-XR has lower maximum delta error than EPA-ng and APPLES, and APPLES has the highest maximum error (specifically, pplacer-XR has maximum

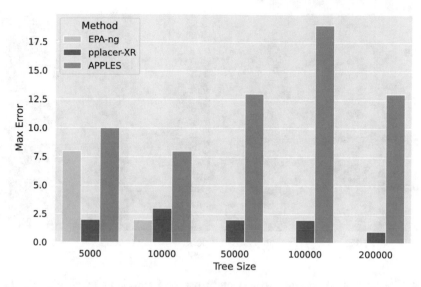

Fig. 3. Maximum delta error over of all placements for each method over five backbone tree sizes. Averages for the 5,000, 10,000, 50,000, and 100,000 taxa trees are over all 1,000 query sequence placements, and for the 200,000 taxa tree the averages are for 200 query sequence placements. EPA-ng was unable to run for trees larger than 10,000 taxa and so no results are reported for EPA-ng on those datasets. This was run on the UIUC campus cluster with 64 GB of memory per placement and 4 h time limit.

error of 3, while EPA-ng and APPLES have a maximum delta error of 8 and 19 respectively.

Finally, Fig. 4 and Table 1 shows the fraction of times that the query is optimally placed (so that delta error is 0); here, too, pplacer-XR outperforms APPLES and EPA-ng, with APPLES having poorer results than EPA-ng.

Table 1. Fraction optimally placed query sequences on backbone trees of size n

	$n = 5000$	$n = 10,000$	$n = 50,000$	$n = 100,000$	$n = 200,000$
pplacer-XR	0.858	0.872	0.919	0.918	0.925
EPA-ng	0.781	0.787	X	X	X
APPLES	0.770	0.785	0.846	0.838	0.875

4.2 Time Analysis

One advantage of distance-based APPLES over maximum likelihood placement methods is generally improved runtime [2]. This can be seen in the results from Fig. 5 and Table 4. For the trees with 5,000 taxa, pplacer-XR is able to place

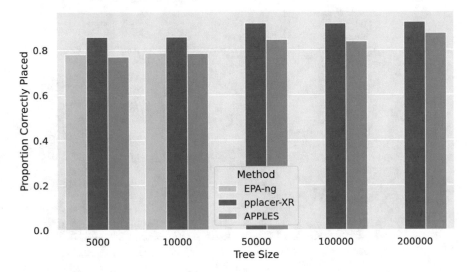

Fig. 4. Fraction of placements with a delta error of zero over of all placements for each method over five backbone tree sizes. Averages for the 5,000, 10,000, 50,000, and 100,000 taxa trees are over all 1,000 query sequence placements, and for the 200,000 taxa tree the averages are for 200 query sequence placements. EPA-ng was unable to run for trees larger than 10,000 taxa and so no results are reported for EPA-ng on those datasets. This was run on the UIUC campus cluster with 64 GB of memory per placement and 4-h time limit.

in half the time of EPA-ng, but takes approximately ten times as much time as APPLES. On trees of 10,000 taxa, the runtime difference between EPA-ng and pplacer-XR grows further to over 3 times the speed, but pplacer-XR is now just 8.3 times slower than APPLES. On the largest tree of 200,000 taxa, pplacer-XR is only 2.7 times slower than APPLES, suggesting the significant speed advantage of APPLES (and potentially other distance-based methods) degrades as the backbone tree size increases. However, on these tree sizes, pplacer-XR was never as fast as APPLES. Finally, EPA-ng was not able to run on the 200,000-taxon backbone tree due to memory limitations and is not shown.

Table 2. Average delta error (Δe) in backbone trees of size n

	$n = 5000$	$n = 10,000$	$n = 50,000$	$n = 100,000$	$n = 200,000$
			Δe		
pplacer-XR	0.150	0.132	0.085	0.084	0.075
EPA-ng	0.239	0.219	X	X	X
APPLES	0.366	0.330	0.239	0.247	0.250

Table 3. Maximum delta error (e_{Max}) on backbone trees of size n

	$n = 5000$	$n = 10{,}000$	$n = 50{,}000$	$n = 100{,}000$	$n = 200{,}000$
		e_{Max}			
pplacer-XR	2	3	2	2	1
EPA-ng	8	2	X	X	X
APPLES	10	8	13	19	13

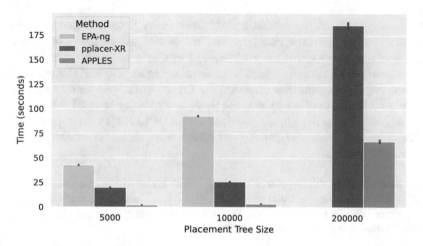

Fig. 5. Average time of all placements for each method over three backbone tree sizes. Averages for the 5,000- and 10,000-taxon trees are over all 1,000 query sequence placements, and for the 200,000-taxon tree the averages are for 200 query sequence placements. EPA-ng was unable to run for trees with 200,000 taxa. This was run on the UIUC campus cluster with 64 GB of memory per placement and 4 h time limit.

Table 4. Average placement time in seconds on backbone trees of size n

	$n = 5000$	$n = 10{,}000$	$n = 200{,}000$
pplacer-XR	20	25	185
EPA-ng	42	92	X
APPLES	2	3	67

5   Conclusions

We have presented pplacer-XR, a technique that uses pplacer, the most accurate placement method, as a step inside a three-stage process for phylogenetic placement. As this study shows, pplacer-XR matches the scalability of APPLES (to our knowledge, the only previous placement method able to perform phylogenetic placement on ultra-large backbone trees with 200,000 leaves), but improves on its accuracy. We also show that pplacer-XR improves on the accuracy, speed, and scalability of EPA-ng, the previous most scalable likelihood-based placement method. Thus, pplacer-XR is an advance in phylogenetic placement methods for large placement trees.

This study suggests several directions for future work. The main advantage of APPLES over pplacer-XR is speed, where APPLES is definitely better (roughly half to one-third the running time of pplacer-XR). Thus, future work should examine ways to reduce the running time for pplacer-XR. It is also possible that changes to the algorithmic design (e.g., how the placement subtree is defined) could improve the accuracy of pplacer-XR. A major limitation of this study is that we only explored phylogenetic placement when given the true alignment of the query sequences to the backbone sequences and where the query sequences were full length; hence, future work should evaluate the case where the input query sequences are short (perhaps with sequencing errors) and unaligned to the backbone sequences. Another direction for future work is to modify pplacer-XR to handle multiple queries at the same time, a capability that EPA-ng is specifically designed to handle. Finally, we note that APPLES2 [1], an improvement on APPLES, has just been developed, and future work should compare pplacer-XR to APPLES2.

Acknowledgments. The research presented here is the result of a course project by EW and YC for the Spring 2020 course CS 581: Algorithmic Genomic Biology, at the University of Illinois, taught by TW. This work was supported in part by the National Science Foundation grant ABI-1458652 to TW.

Appendix

Commands to create backbone trees: All placement methods use the same backbone tree topologies, but have different branch lengths (following the protocol as provided in [2]). We downloaded the backbone trees with their optimized branch lengths for each phylogenetic placement method from the APPLES repository.

On replicates 1 and 2 of the 100,000-leaf backbone condition and replicate 0 of the 200,000-leaf replicate backbone condition, each containing more than two identical sequences, the backbone trees had polytomies. We randomly resolved these in order to run RAxML.

Random tree refinement: raxmlHPC-PTHREADS -f e -t res_true.fasttree -m GTRGAMMA -s aln_dna.phy -n REF -p 1984 -T 16

APPLES command: run_apples.py -t backbone.tree -s ref.fa -q query.fa -T 16 -o apples.jplace

EPA-ng command: epa-ng –ref-msa ref.fa –tree backbone.tree –query query.fa –outdir $query –model RAxML_info.REF8 –redo -T 16

pplacer-XR commands: python3 pplacer-XR.py GTR RAxML_info.REF backbone.tree output_dir aln.fa query.txt 2000

Github site: https://github.com/chry04/pplacer_plusplus

References

1. Balaban, M., Roush, D., Zhu, Q., Mirarab, S.: APPLES-2: faster and more accurate distance-based phylogenetic placement using divide and conquer. bioRxiv (2021). https://doi.org/10.1101/2021.02.14.431150
2. Balaban, M., Sarmashghi, S., Mirarab, S.: APPLES: scalable distance-based phylogenetic placement with or without alignments. Syst. Biol. **69**(3), 566–578 (2020)
3. Barbera, P., et al.: EPA-NG: massively parallel evolutionary placement of genetic sequences. Syst. Biol. **68**(2), 365–369 (2019)
4. Berger, S.A., Krompass, D., Stamatakis, A.: Performance, accuracy, and web server for evolutionary placement of short sequence reads under maximum likelihood. Syst. Biol. **60**(3), 291–302 (2011)
5. Bik, H.M., Porazinska, D.L., Creer, S., Caporaso, J.G., Knight, R., Thomas, W.K.: Sequencing our way towards understanding global eukaryotic biodiversity. Trends Ecol. Evol. **27**(4), 233–243 (2012)
6. Chaumeil, P.A., Mussig, A.J., Hugenholtz, P., Parks, D.H.: GTDB-Tk: a toolkit to classify genomes with the genome taxonomy database. Bioinformatics **36**(6), 1925–1927 (2020)
7. Conlan, S., Kong, H.H., Segre, J.A.: Species-level analysis of DNA sequence data from the NIH Human Microbiome Project. PLoS ONE **7**(10), e47075 (2012)
8. Matsen, F.A., Kodner, R.B., Armbrust, E.V.: pplacer: linear time maximum-likelihood and Bayesian phylogenetic placement of sequences onto a fixed reference tree. BMC Bioinform. **11**(1), 538 (2010)
9. McCoy, C.O., Matsen IV, F.A.: Abundance-weighted phylogenetic diversity measures distinguish microbial community states and are robust to sampling depth. PeerJ **1**, e157 (2013)
10. Mirarab, S., Nguyen, N., Guo, S., Wang, L.S., Kim, J., Warnow, T.: PASTA: ultra-large multiple sequence alignment for nucleotide and amino-acid sequences. J. Comput. Biol. **22**(5), 377–386 (2015)
11. Mirarab, S., Nguyen, N., Warnow, T.: SEPP: SATé-enabled phylogenetic placement. In: Biocomputing 2012, pp. 247–258. World Scientific (2012)
12. Nguyen, N.P., Mirarab, S., Liu, B., Pop, M., Warnow, T.: TIPP: taxonomic identification and phylogenetic profiling. Bioinformatics **30**(24), 3548–3555 (2014)
13. Price, M.N., Dehal, P.S., Arkin, A.P.: FastTree 2-approximately maximum-likelihood trees for large alignments. PLoS ONE **5**(3), e9490 (2010)
14. Shah, N., Molloy, E.K., Pop, M., Warnow, T.: TIPP2: metagenomic taxonomic profiling using phylogenetic markers. Bioinformatics (2021)
15. Stamatakis, A.: RAxML-VI-HPC: maximum likelihood-based phylogenetic analyses with thousands of taxa and mixed models. Bioinformatics **22**(21), 2688–2690 (2006)
16. Tavaré, S.: Some probabilistic and statistical problems in the analysis of DNA sequences. Lect. Math. Life Sci. **17**(2), 57–86 (1986)
17. Yang, Z.: Maximum likelihood phylogenetic estimation from DNA sequences with variable rates over sites: approximate methods. J. Mol. Evol. **39**(3), 306–314 (1994)

Comparing Methods for Species Tree Estimation with Gene Duplication and Loss

James Willson⬤, Mrinmoy Saha Roddur⬤, and Tandy Warnow[✉]⬤

Department of Computer Science, University of Illinois at Urbana-Champaign,
Urbana, IL, USA
{jamessw3,mroddur2,warnow}@illinois.edu

Abstract. Species tree inference from gene trees is an important part of biological research. One confounding factor in estimating species trees is gene duplication and loss, which can lead to gene family trees with multiple copies of the same species. In recent years there have been several new methods developed to address this problem that have substantially improved on earlier methods; however, the best performing methods (ASTRAL-Pro, ASTRID-multi, and FastMulRFS) have not yet been directly compared. In this study, we compare ASTRAL-Pro, ASTRID-multi, and FastMulRFS under a wide variety of conditions. Our study shows that while all three have nearly the same accuracy under most conditions, ASTRAL-Pro and ASTRID-multi are more reliably accurate than FastMuLRFS (with a small advantage to ASTRID-multi), and that ASTRID-multi is often faster than ASTRAL-Pro.

Keywords: Species-tree inference · Gene duplication and loss

1 Introduction

Species tree estimation is an important part of biological research. One of the main challenges is gene tree heterogeneity caused by phenomena such as gene duplication and loss (GDL), incomplete lineage sorting (ILS), horizontal gene transfer, and hybrid speciation. When these events occur, the tree for a given locus (i.e., "gene tree" or "gene family tree") may be different from the species tree, making the inference of the species tree very challenging. However, the reconciliation of species trees and gene family trees can yield insight into biological function and processes [6,10,11], making these tree inferences important. There are many methods available to estimate species trees in the presence of ILS that are proven statistically consistent under the Multi-Species Coalescent [13] model; see [16] for a 2015 review, and [12,24] for two recent methods. However GDL is

---

J. Willson and M. S. Roddur—Equal contribution.

C. Martín-Vide et al. (Eds.): AlCoB 2021, LNBI 12715, pp. 106–117, 2021.
https://doi.org/10.1007/978-3-030-74432-8_8

another of the primary causes of gene tree heterogeneity, and so methods that can estimate species trees with high accuracy in the presence of GDL are needed.

One approach to species tree estimation in the presence of GDL presumes the ability to identify orthologs (i.e., genes in different species that have evolved from a common ancestor via a speciation rather than a duplication event). Given the accurate identification of orthologs, species tree estimation can proceed by restricting analyses to single-copy gene trees. However, orthology detection is not yet reliably solved [9], and so this approach can result in errors in the input (i.e., the single copy gene trees may not reflect the species tree), and hence errors in the output species tree. Furthermore, restricting to orthologous genes reduces the amount of available phylogenetic signal, and so can reduce accuracy.

An alternative approach is to use the multi-copy genes (either their alignments or their gene family trees, called "MUL-trees"), and then use these to estimate the species tree. Perhaps the most well known approaches are based on gene tree parsimony, which seek to minimize the total number of duplications and losses; examples include DupTree [28], iGTP [3], and DynaDup [1]. Other methods include PHYLDOG [2] and guenomu [20], which use probabilistic models of gene evolution, STAG [7], and MulRF [4]. None of these methods have yet been proven to be statistically consistent under GDL models.

In 2019, a species tree estimation method that was developed for the ILS-only scenario, ASTRAL-multi [22], was proven to be statistically consistent under GDL [15]. Subsequently, ASTRAL-Pro [29] was developed to explicitly address GDL; its technique requires "rooting and tagging" each input gene tree, and if the rooting and tagging is correctly performed, then ASTRAL-Pro is statistically consistent under GDL. Interestingly, although ASTRAL-Pro has not been proven statistically consistent under GDL without the assumption of proper tagging, it has been shown to be more accurate than ASTRAL-multi in extensive simulation studies [29].

Two other methods of interest are ASTRID-multi and FastMulRFS. ASTRID-multi (available at [26]) is an extension by Pranjal Vachaspati of the ASTRID [27] method for ILS-based species tree estimation to allow it work with multiple individuals. FastMulRFS [18], like MulRF, is a method for the Robinson-Foulds Supertree problem adapted to the MUL-tree setting. Both FastMulRFS and ASTRID-multi have been found to be more accurate than ASTRAL-multi in simulation studies evaluating methods for species tree estimation under models that include GDL and varying levels of ILS [15,18,29]. However, a comparison between these three leading methods has not been performed, and no analyses have been made of these methods when gene trees are missing species (i.e., the "missing data" condition), which is very common in biological datasets.

Here, we report on a study comparing ASTRAL-Pro, ASTRID-multi, and FastMulRFS in terms of topological accuracy and running time on data sets we simulated under the DLCOAL model [23] that includes GDL and ILS. We explore performance under a wide range of model conditions, including large numbers of species (up to 1000) and genes (up to 10,000). Furthermore, unlike previous studies, we include conditions where gene trees are "incomplete", and so are missing species. We find that differences in accuracy between the three methods tend to be small, but that there are conditions where FastMulRFS is not quite as accurate as either ASTRAL-Pro or ASTRID-multi. The comparison between ASTRID-multi and ASTRAL-Pro shows very close accuracy under most conditions (with a slight edge to ASTRAL-Pro) but that ASTRID-multi is faster. Finally, all three methods succeeded in analyzing datasets with 1000 species and 1000 genes that evolved with GDL and ILS, and so can analyze very large datasets.

2   Experimental Design

To compare ASTRAL-Pro, FastMulRFS, and ASTRID-multi, we designed a simulation study with three experiments, each containing both GDL and ILS. To ensure reproducibility, the datasets generated for this study are freely available in the Illinois Data Bank at https://databank.illinois.edu/datasets/IDB-2418574. Commands necessary to reproduce the experiment are provided in the Supplementary Materials, provided at http://tandy.cs.illinois.edu/gdl-suppl.pdf.

All three methods run in polynomial time, but their running times depend on the input properties, as we now describe. FastMuLRFS and ASTRAL-Pro construct a constraint set of allowed bipartitions, which includes at a minimum all the true bipartitions from the input gene trees, and the running times for their dynamic programming algorithms (to find optimal solutions within the constrained search space) runs nearly quadratically in the size of the constraint space. In contrast, ASTRID-multi computes a distance matrix, and then computes a tree on the distance matrix using FastME [14] (if there is no missing data in the matrix) or BioNJ* [5] if the distance matrix has some missing data. We designed our study to explore conditions that might affect these methods differently. In Experiment 1, we examine how the methods perform on complete gene trees (i.e., no missing data) under various conditions. In Experiment 2, we examine the methods on gene family trees where some gene trees can be incomplete (i.e., the missing data condition). Finally, in Experiment 3: we evaluate the methods scalability by running the methods on a simulated data set with 1000 species.

The simulation study evolved gene trees within species trees using SimPhy [17], and then evolved sequences within the gene trees using INDELIBLE [8] under the GTRGAMMA model of sequence evolution. Gene trees were then estimated on these sequence alignments using FastTree2 [21], a fast maximum likelihood heuristic, and provided as input to the different species tree estimation methods. We varied the number of genes, ILS level, GDL level, the relative probabilities of gene duplications to losses, and the sequence length per gene. Because we use simulated data, the true species tree and true gene trees are known. This allows us to exactly quantify error in both the estimated gene trees and the estimated species trees. For these error calculations, we report the Robinson-Foulds [25] error rate, which is the fraction of the set of bipartitions (defined for the estimated and true trees) that only appear in one of the two trees; hence, the Robinson-Foulds error varies between 0 (identical trees) and 1 (no bipartitions in common). Mean gene tree estimation error (MGTE) affects species tree estimation methods that combine gene trees [15,18,29], and so we vary the sequence length per gene to explore this impact in our study.

For Experiments 1 and 2, we ran the methods on the University of Illinois at Urbana-Champaign Campus Cluster, which imposes a four hour restriction and has 64 Gb of memory available (at a minimum); for Experiment 3 we use the Tallis queue which allows four-week analyses and has 256 Gb of available memory. In each experiment, we note which methods failed and for what reason (e.g., failure to complete within the allowed time or insufficient memory).

In our simulations, we varied the number of species (from 100 to 1000), the number of genes (from 100 to 10,000), the degree of ILS, the GDL rate, the relative frequency of duplications and losses, and the mean gene tree estimation error. While varying a given condition, we kept the rest of the conditions locked to the following default values: 100 species, 1000 or 10,000 gene trees estimated from 100 bp alignments, haploid effective population size of 5.0×10^7, duplication rate 5.0×10^{-10}, and a loss rate equal to the duplication rate. Under the default conditions, the mean gene tree estimation error (MGTE) was moderate (43% MGTE) and the ILS levels were also moderate (average topological distance, denoted AD, between true gene trees and true species trees of 20%). The empirical statistics for these datasets are available in Table S2 in the Supplementary Materials.

Experiment 1: We simulated a set of 100-taxon data sets with varying properties. We tested data sets with 100, 500, 1000, and 10,000 gene trees. We chose duplication rates of 1.0×10^{-10}, 5.0×10^{-10}, and 1.0×10^{-9}, all run with relative loss rates of 0, 0.5, and 1, giving us a total of nine conditions. As seen in Table 1, this gives us a set of model conditions that vary substantially in terms of average number of leaves in the resulting gene family trees, ranging from 117 leaves to 3728 leaves. We explore results using both true and estimated gene trees, and varied the sequence length per gene from 100 to 500 bp; this produced datasets with mean gene tree error (MGTE) of 43% and 19.2%, respectively. We controlled the level of ILS through the effective haploid population size, choosing sizes of 1.0×10^4, 5.0×10^7, and 2.0×10^8. These produce true gene trees that have average topological distance (AD) using normalized RF distance to the true species tree of 0.000468%, 20.3019%, and 50.00392%, respectively; we round these values to 0% (low ILS), 20% (moderate ILS) and 50% (high ILS).

Table 1. Average number of leaves in the true gene family trees for Experiment 1 (100 taxa), for different duplication rates (rows) and ratios of losses to duplications (columns). Results shown are averaged across 1000 genes and replicates.

	L/D = 0	L/D = 0.5	L/D = 1
1×10^{-10}	145.1	128.0	116.6
5×10^{-10}	550.0	290.6	165.3
1×10^{-9}	3727.8	993.0	228.5

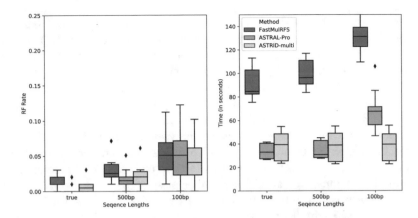

Fig. 1. *Experiment 1.* Impact of mean gene tree estimation error (MGTE) on species tree error (RF error rates) and wall clock running time (seconds); averages over 10 replicates per model condition are shown. All the data sets have 100 species, 1000 gene trees, AD = 20%, a duplication rate of 5.0×10^{-10} and an equal loss rate. The data sets include true gene trees and estimated gene trees from two sequence lengths, and have MGTE of 0%, 19.2%, and 0.43.2%, respectively. The boxes stretch from the 1st to 3rd quartile and the lines through the boxes shows the median.

Experiment 2: We generated datasets with missing data under the M_{clade} model of missing data [19]: We listed all the clades from the species tree containing at least 20% of all the leaves in the tree. For each gene family tree, we randomly picked one of these species tree clades and deleted all the leaves from that gene family tree not in the selected clade. On average, this protocol deleted 40% of the species from the gene trees. We ran this process on the 100-taxon data set from Experiment 1 with 100, 500, and 1000 estimated gene trees.

Experiment 3: We simulated an additional data set with 1000 species, where the rest of the conditions were identical to the default conditions in Experiment 1. This experiment was run on the Tallis cluster, which allows the methods an unlimited amount of time to run and extra available memory.

3   Results

Experiment 1: Our first analysis examined the impact of changing the gene tree estimation error on species tree accuracy (Fig. 1). Given true gene trees, ASTRAL-Pro produced the true species tree while the other methods had some error (but very low error, at most 2%). On estimated gene trees, species tree error rates increased for all methods, with FastMulRFS slightly worse than the other two methods and no statistically significant difference between ASTRAL-Pro and ASTRID-multi. For running time, FastMulRFS had the highest running time, followed by ASTRAL-Pro, and then by ASTRID-multi. Furthermore, ASTRID-multi's running time seemed completely unaffected by the change in MGTE, but FastMulRFS and ASTRAL-Pro both increased in running time as MGTE increased.

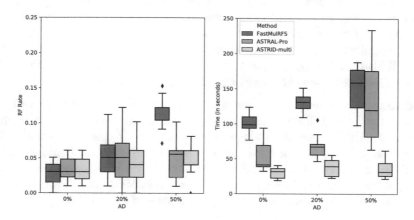

Fig. 2. *Experiment 1.* Impact of ILS level on tree error (RF error rates) and wall clock running time (seconds); averages across 10 replicates per model condition are shown. All the data sets have 100 species, 1000 gene trees estimated from 100 bp alignments (34.7%, 43.2%, 43.9% MGTE), a duplication rate of 5.0×10^{-10}, and an equal loss rate.

The results on the data sets with varying levels of ILS (Fig. 2) show that increasing ILS results in increased error for all methods, but the methods are comparable in accuracy until the high ILS condition (AD = 50%), where Fast-MulRFS has a substantial increase in error rates, while the other methods remain fairly low in error. The running time shows an interesting trend: as the amount of ILS increases the running times for all methods increase, except for ASTRID-multi which remains unaffected. Also, FastMulRFS is the slowest of the methods, ASTRAL-Pro is intermediate, and ASTRID-multi is the fastest.

We then examined the results for varying the duplication and loss rates (Fig. 3). For the most part, the methods stayed competitive at lower duplication rates, but as the duplication rates increased FastMulRFS's error rate increased, making it clearly less accurate than the others; this trend held for all duplication/loss ratios tested. ASTRID-multi was very slightly, yet statistically significantly, less accurate then ASTRAL-Pro under the highest duplication rate as well ($p = 0.0282$). The running time revealed interesting trends. For the lowest duplication rate, ASTRID-multi was the fastest and FastMulRFS the slowest, but at the highest duplication rate they reversed positions.

Next we examined the data sets which varied the number of gene trees (Fig. 4). All methods increased in accuracy and running time as the number of gene trees increase. The differences between methods are largest at 100 genes, where FastMulRFS has lower accuracy than the other methods, and then reduce as the number of genes increase, so that by 1000 genes, the differences are minor. Running time, however, is substantially impacted by the number of genes, with

ASTRID-multi always faster than the other methods and FastMulRFS most often the slowest.

Experiment 2: Next we investigated the effects of missing data (Fig. 5). Results shown with missing data are very similar to results shown without missing data (i.e., all gene trees have all the species); with few genes there is a slight, but statistically significant ($p = 0.0446$), decrease in ASTRID-multi's accuracy relative to ASTRAL-Pro. The performance with respect to running time is also similar, with ASTRID-multi fastest and FastMulRFS slowest overall.

Experiment 3: We compared methods when analyzing 1000-species 1000-gene datasets, using the Tallis cluster, which allows unlimited time and 256 Gb memory (Fig. 6). Differences on the five replicates that all methods completed on are very small (and were not statistically significant). ASTRID-multi had the advantage when it came to running time, though ASTRAL-Pro was not too much worse (neither exceeded 3 h for the majority of replicates), and FastMulRFS was by far the slowest with a median time to complete of around 8 h.

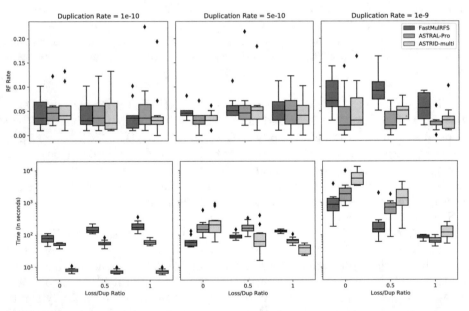

Fig. 3. *Experiment 1.* Impact of GDL rate and loss/dup rate on species tree error (RF error rates) and wall clock running time (seconds); averages across 10 replicates are shown. All the data sets have 100 species, 1000 gene trees, gene trees estimated from 100 bp alignments (44.1%, 44.9%, 44.0%, 43.1%, 45.1%, 43.2%, 47.8%, 43.5%, 40.0% MGTE), and AD = 20%. ASTRAL-Pro and ASTRID-multi failed on some replicates; results shown here are for the replicates on which all methods completed.

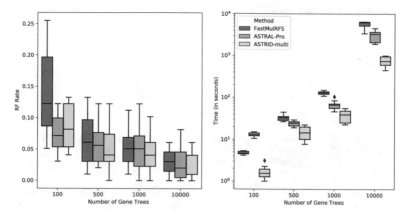

Fig. 4. *Experiment 1.* Impact of number of genes on species tree error (RF error rates) and wall clock running time (seconds); results shown are averaged across 10 replicates. All the data sets have 100 species, gene trees estimated from 100 bp alignments (43.1%, 43.3%, 43.3%, 41.2% MGTE), AD = 20%, a duplication rate of 5.0×10^{-10} and an equal loss rate. ASTRAL-Pro ran out of memory on some of the 10,000 gene tree replicates; results shown here are for the replicates on which all methods completed.

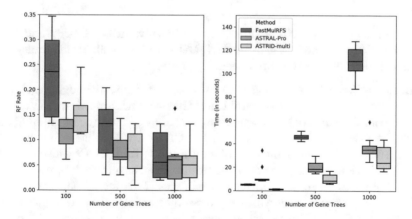

Fig. 5. *Experiment 2.* Impact of number of gene trees on species tree error (RF error rates) and wall clock running time (seconds) under the missing data condition; results shown are averaged across 10 replicates. All datasets contain 100 species and gene trees estimated from 100 bp alignments (43.1%, 43.3%, 43.2% MGTE). All the data sets have AD = 20%, a duplication rate of 5.0×10^{-10}, and an equal loss rate.

Statistical Significance: To test for statistical significance between the results for ASTRID-multi and ASTRAL-Pro (both of which yielded very similar results), we ran a paired t-test between the results on most of our module conditions (Supplementary Materials, Table S3). We found that the only cases in which there was a statistically significant difference between the accuracy of these two methods was under the highest duplication rate (1×10^{-9}) and also the lowest

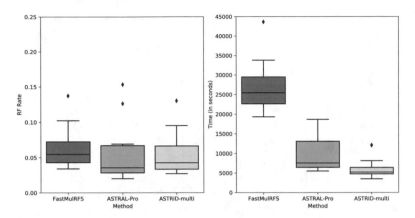

Fig. 6. *Experiment 3.* Species tree error (RF error rates) and wall clock running time (seconds) on the dataset containing 1000 species and 1000 gene trees estimated from 100 bp alignments (44.4% MGTE). All the data sets have AD = 20%, a duplication rate of 5.0×10^{-10}, and an equal loss rate. Results shown are averaged across 10 replicates. All methods were run on Tallis, which has 256 Gb available memory and allows up to 4 weeks of analysis.

number of gene trees (100) with missing data (p-values of 0.0282 and 0.0446, respectively). In both of these cases ASTRAL-Pro had a slight advantage. More details can be found in the supplementary material.

Failures to Complete: During the course of the study, although FastMulRFS always completed, there were several times that ASTRID-multi or ASTRAL-Pro failed to complete due either to exceeding the allowed runtime or running out of memory. All these occurred on the Campus Cluster, which has a four hour time limit and 64 Gb of available memory at a minimum (Table 2). (We also noted that ASTRAL-Pro had difficulty calculating branch lengths on the datasets with 1000 species, failing to do so on 4 of the 10 replicates. This forced us to re-run these replicates using an argument which suppressed branch length and branch support estimation.) ASTRID-multi crashed twice due to memory constraints (once on a replicate with 10,000 gene trees and once on a replicate with a high duplication rate) and timed out on two high duplication rate replicates. ASTRAL-Pro timed out on five different replicates (one high duplication rate replicate and four 10,000 gene tree replicates) and crashed on five more due to lack of memory (again, one high duplication rate replicate and four 10,000 gene tree replicates).

ASTRID-multi runs out of memory when its distance matrix becomes too large, which occurs when the total number of copies of all genes across all species becomes large. The reason that ASTRAL-Pro runs out of memory or exceeds the allowed running time is that its constraint set becomes too large under some conditions (which affects memory usage) and its running time is almost quadratic in the constraint set size. Furthermore, the constraint set becomes large when there are many genes with substantial discord between them due to either gene tree estimation error, high ILS, or high GDL rates. Thus, high

Table 2. Failures on the Campus Cluster (4 h time limit). "Total Failures" includes the total of all failures including out of memory errors (OOM) and timeouts. The campus cluster is heterogeneous; however, at a minimum it has 64 GB of memory. The numbers after GDL is the duplication and loss rate ratio respectively and the number after 10,000 gene trees is the loss rate ratio. FastMulRFS is not included as it did not fail to complete in these analyses.

Method	Experiment	Total failures	OOM	Timeouts
ASTRAL-Pro	GDL $(1.0 \times 10^{-9}/0)$	20%	10%	10%
	10,000 gene trees (0)	10%	10%	0%
	10,000 gene trees (0.5)	40%	0%	40%
	10,000 gene trees (1)	30%	30%	0%
ASTRID-multi	GDL $(1.0 \times 10^{-9}/0)$	30%	10%	20%
	10,000 gene trees (0)	20%	10%	0%

gene duplication rates, especially with large numbers of species, is a problem for ASTRID's memory usage, while other factors (especially large numbers of genes and species, high gene tree estimation error, and high ILS) are problematic for ASTRAL-Pro's memory usage and running time.

4   Discussion and Conclusion

The three methods we compared, ASTRID-multi, ASTRAL-Pro, and FastMul-RFS, were generally close in accuracy under most conditions, but even so some differences can be discerned. Most importantly, there were no model conditions where FastMulRFS was more accurate than the other methods. Furthermore, when there were substantial differences in accuracy (which occurred with high ILS, high duplication rates, or small numbers of genes), FastMuLRFS had higher error rates than the other methods. The comparison in accuracy between ASTRID-multi and ASTRAL-Pro is therefore of greater interest. In nearly every model condition, there was no statistically significant difference between the accuracy of the methods; however, small numbers of genes with missing data tended to slightly favor ASTRAL-Pro, as well as high duplication rates.

However, there are differences in terms of computational requirements. Under most conditions, ASTRID-multi is much faster than both ASTRAL-Pro and FastMulRFS, especially for the more challenging datasets where the overall gene tree heterogeneity (which is impacted by MGTE, ILS level, and number of genes) is high. The explanation for this trend is that both FastMuLRFS and ASTRAL-Pro construct a constraint set of allowed bipartitions, which becomes large under some conditions, which can make them require excessive memory and running time. However, ASTRID-multi computes a distance matrix and then computes a tree on the distance matrix; this is fast except when the duplication rate is very high, and is not impacted by MGTE or ILS (and only minimally impacted by the number of genes). Hence, in nearly all cases, except where there is a very high duplication rate, ASTRID-multi is the fastest of these methods.

Given the close results in terms of accuracy under most conditions, the choice of method to use may depend on dataset size and computational resources. When memory usage and running time are not of concern, all three methods could be used, and the bipartitions that appear in all the trees can be considered reliable. When computational resources are more limited, then both ASTRID-multi and ASTRAL-Pro can be used. Finally, if running time or memory usage is a significant issue, then ASTRID-multi may be an acceptable choice.

Acknowledgments. This work made use of the Illinois Campus Cluster, a computing resource that is operated by the Illinois Campus Cluster Program (ICCP) in conjunction with the National Center for Supercomputing Applications (NCSA) and which is supported by funds from the University of Illinois at Urbana-Champaign.

References

1. Bayzid, M.S., Mirarab, S., Warnow, T.: Inferring optimal species trees under gene duplication and loss. In: Biocomputing 2013, pp. 250–261. World Scientific (2013)
2. Boussau, B., Szöllősi, G.J., Duret, L., Gouy, M., Tannier, E., Daubin, V.: Genome-scale coestimation of species and gene trees. Genome Res. **23**(2), 323–330 (2013)
3. Chaudhary, R., Bansal, M.S., Wehe, A., Fernández-Baca, D., Eulenstein, O.: iGTP: a software package for large-scale gene tree parsimony analysis. BMC Bioinform. **11**(1), 1–7 (2010)
4. Chaudhary, R., Fernández-Baca, D., Burleigh, J.G.: MulRF: a software package for phylogenetic analysis using multi-copy gene trees. Bioinformatics **31**(3), 432–433 (2015)
5. Criscuolo, A., Gascuel, O.: Fast NJ-like algorithms to deal with incomplete distance matrices. BMC Bioinform. **9**(1), 1–16 (2008). https://doi.org/10.1186/1471-2105-9-166
6. Dittmar, K., Liberles, D.: Evolution After Gene Duplication. Wiley, Hoboken (2011)
7. Emms, D., Kelly, S.: STAG: species tree inference from all genes, p. 267914. BioRxiv (2018)
8. Fletcher, W., Yang, Z.: INDELible: a flexible simulator of biological sequence evolution. Mol. Biol. Evol. **26**(8), 1879–1888 (2009)
9. Glover, N., et al.: Advances and applications in the Quest for Orthologs. Mol. Biol. Evol. **36**(10), 2157–2164 (2019)
10. Goodman, M., Czelusniak, J., Moore, G.W., Romero-Herrera, A.E., Matsuda, G.: Fitting the gene lineage into its species lineage, a parsimony strategy illustrated by cladograms constructed from globin sequences. Syst. Biol. **28**(2), 132–163 (1979)
11. Innan, H., Kondrashov, F.: The evolution of gene duplications: classifying and distinguishing between models. Nat. Rev. Genet. **11**(2), 97–108 (2010)
12. Kim, A., Degnan, J.H.: PRANC: ML species tree estimation from the ranked gene trees under coalescence. Bioinformatics **36**(18), 4819–4821 (2020)
13. Kingman, J.F.C.: The coalescent. Stochast Process. Appl. **13**(3), 235–248 (1982)
14. Lefort, V., Desper, R., Gascuel, O.: FastME 2.0: a comprehensive, accurate, and fast distance-based phylogeny inference program. Mol. Biol. Evol. **32**(10), 2798–2800 (2015)

15. Legried, B., Molloy, E.K., Warnow, T., Roch, S.: Polynomial-time statistical estimation of species trees under gene duplication and loss. In: Schwartz, R. (ed.) RECOMB 2020. LNCS, vol. 12074, pp. 120–135. Springer, Cham (2020). https://doi.org/10.1007/978-3-030-45257-5_8
16. Liu, L., Wu, S., Yu, L.: Coalescent methods for estimating species trees from phylogenomic data. J. Syst. Evol. **53**(5), 380–390 (2015)
17. Mallo, D., de Oliveira Martins, L., Posada, D.: SimPhy: phylogenomic simulation of gene, locus, and species trees. Syst. Biol. **65**(2), 334–344 (2016)
18. Molloy, E.K., Warnow, T.: FastMulRFS: fast and accurate species tree estimation under generic gene duplication and loss models. Bioinformatics **36**(Supplement_1), i57–i65 (2020)
19. Nute, M., Chou, J., Molloy, E.K., Warnow, T.: The performance of coalescent-based species tree estimation methods under models of missing data. BMC Genomics **19**(5), 1–22 (2018)
20. de Oliveira Martins, L., Posada, D.: Species tree estimation from genome-wide data with Guenomu. In: Keith, J.M. (ed.) Bioinformatics. MMB, vol. 1525, pp. 461–478. Springer, New York (2017). https://doi.org/10.1007/978-1-4939-6622-6_18
21. Price, M.N., Dehal, P.S., Arkin, A.P.: FastTree 2-approximately maximum-likelihood trees for large alignments. PLoS ONE **5**(3), e9490 (2010)
22. Rabiee, M., Sayyari, E., Mirarab, S.: Multi-allele species reconstruction using ASTRAL. Mol. Phylogenet. Evol. **130**, 286–296 (2019)
23. Rasmussen, M.D., Kellis, M.: Unified modeling of gene duplication, loss, and coalescence using a locus tree. Genome Res. **22**(4), 755–765 (2012)
24. Richards, A., Kubatko, L.: Bayesian weighted triplet and quartet methods for species tree inference. arXiv (2020)
25. Robinson, D.F., Foulds, L.R.: Comparison of phylogenetic trees. Math. Biosci. **53**(1–2), 131–147 (1981)
26. Vachaspati, P.: ASTRID (2018–2021). https://github.com/pranjalv123/ASTRID
27. Vachaspati, P., Warnow, T.: ASTRID: accurate species trees from internode distances. BMC Genomics **16**(S10) (2015). Article number: S3. https://doi.org/10.1186/1471-2164-16-S10-S3
28. Wehe, A., Bansal, M.S., Burleigh, J.G., Eulenstein, O.: DupTree: a program for large-scale phylogenetic analyses using gene tree parsimony. Bioinformatics **24**(13), 1540–1541 (2008)
29. Zhang, C., Scornavacca, C., Molloy, E.K., Mirarab, S.: ASTRAL-Pro: quartet-based species-tree inference despite Paralogy. Mol. Biol. Evol. **37**(11), 3292–3307 (2020)

Sequence Alignment and Genome Rearrangement

Reversal Distance on Genomes with Different Gene Content and Intergenic Regions Information

Alexsandro Oliveira Alexandrino[1](✉) , Klairton Lima Brito[1] ,
Andre Rodrigues Oliveira[1] , Ulisses Dias[2] , and Zanoni Dias[1]

[1] Institute of Computing, University of Campinas (Unicamp), Campinas, Brazil
{alexsandro,klairton,andrero,zanoni}@ic.unicamp.br
[2] School of Technology, University of Campinas (Unicamp), Limeira, Brazil
ulisses@ft.unicamp.br

Abstract. Recent works on genome rearrangements have shown that incorporating intergenic region information along with gene order in models provides better estimations for the rearrangement distance than using gene order alone. The reversal distance is one of the main problems in genome rearrangements. It has a polynomial time algorithm when only gene order is used to model genomes, assuming that repeated genes do not exist and that gene orientation is known, even when the genomes have distinct gene sets. The reversal distance is NP-hard and has a 2-approximation algorithm when incorporating intergenic regions. However, the problem has only been studied assuming genomes with the same set of genes. In this work, we consider the variation that incorporates intergenic regions and that allows genomes to have distinct sets of genes, a scenario that leads us to include indels operations (insertions and deletions). We present a 3-approximation algorithm using the labeled intergenic breakpoint graph, which is based on the well-known breakpoint graph structure.

Keywords: Genome rearrangements · Reversals · Indels · Intergenic regions

1 Introduction

In comparative genomics, the computation of the distance between two genomes using genome rearrangements is an important problem with many variations studied in the past decades. A genome rearrangement is a mutational event that modifies the amount of genetic material or changes the position and the orientation of a segment of genetic material in a genome.

The genome rearrangement distance problem tries to compute how many and which mutational events have occurred between two given genomes. Due to the Principle of Parsimony, the most common interpretation for the *rearrangement*

© Springer Nature Switzerland AG 2021
C. Martín-Vide et al. (Eds.): AlCoB 2021, LNBI 12715, pp. 121–133, 2021.
https://doi.org/10.1007/978-3-030-74432-8_9

distance is to define it as the minimum number of rearrangements necessary to transform one genome into another.

The main variations of the genome rearrangement distance arose from (i) the set of rearrangement types considered, which is called the *rearrangement model*, and (ii) how genomes are modeled. Regarding the rearrangement model, we consider *reversals*, which are events that invert the order and the orientation of a segment in a genome, and *indels*, which are events of insertions and deletions of genetic material in a genome.

Concerning how a genome is modeled, seminal works on genome rearrangements considered only gene order for the computation of the distance and assumed that the genomes shared the same set of genes and no gene is repeated [8]. With these assumptions, the sequence of genes in a genome is modeled as a permutation of integer numbers, where each element represents a gene, and the sign of the element represents its orientation. In this case, the rearrangement distance problem is equivalent to the problem of sorting a permutation with a minimum number of rearrangements. The problem of Sorting by Reversals has a polynomial time algorithm for signed permutations (orientation information is available) [9] and it is NP-hard for unsigned permutations (orientation information is not available) [7].

When the genomes do not share the same set of genes, the sequence of genes in each genome is represented by a string. Similarly, each element represents a gene, and the sign of the element indicates its orientation. In this way, indels must be added to the rearrangement model to find a minimum length sequence of rearrangements that transforms one genome into another. The Reversal and Indel Distance problem is also solvable in polynomial time on signed strings [11] and NP-hard on unsigned strings [1].

Recent works incorporated the information about intergenic regions (i.e., sequence of nucleotides between consecutive genes) when modeling genomes. It was recently shown that considering gene order alongside intergenic region sizes can improve the distance estimation in real scenarios [4,5].

When the genomes have the same set of genes, the Reversal Distance Considering Gene Order and Intergenic Regions problem is NP-hard for both signed and unsigned permutations [6,10]. The best results for these problems are a 2-approximation algorithm for signed permutations [10] and a 4-approximation algorithm for unsigned permutations [6]. Indels were added to the rearrangement model in previous works that incorporated intergenic regions, but they were only allowed to affect intergenic regions [6].

In this work, we study the Reversal and Indel Distance Considering Gene Order and Intergenic Regions problem for genomes with distinct gene sets and without duplicated genes. Indels can insert or delete a segment of genomic material containing genes and intergenic regions. We adapted the breakpoint graph [8], which is a well-known structure introduced for the rearrangement distance on permutations, into a structured called labeled intergenic breakpoint graph, in order to model an instance for the problem we are addressing. Then, we present lower bounds for the distance and a 3-approximation algorithm for this problem.

2   Background

The sequence of genes in a genome $\mathcal{G}$ can be represented by a string A, where each element of the string corresponds to a gene, that is, the element A_i corresponds to the i-th gene in the sequence of genes from $\mathcal{G}$. Furthermore, we use a '+' or '−' sign in each element to represent the orientation of the genes.

While it is possible to find an association between the genes in the genomes being compared, this is not possible for intergenic regions since it is known that intergenic regions are more susceptible to change than genes [4]. Therefore, we use a list of non-negative integer numbers $\breve{A}$ to represent the size (number of nucleotides) of the intergenic regions in a genome, such that $\breve{A}_i$ is the size of the intergenic region between genes A_i and A_{i+1}.

In this way, a genome $\mathcal{G}$ is represented by the pair $(A, \breve{A}) = (\breve{A}_1 \ A_1 \ \breve{A}_2 \ A_2 \ \dots \ \breve{A}_n \ A_n \ \breve{A}_{n+1})$, mapping both gene order and intergenic regions sizes.

The sizes of a string A and a list $\breve{A}$ are denoted by $|A|$ and $|\breve{A}|$, respectively. Given a string A, the *alphabet* Σ_A is the set of characters of A without considering signs. We denote by ι^n the identity string of size n that is equal to $(1 \ 2 \ \dots \ n)$.

Since we deal with genomes without duplicated genes, when comparing two genomes $\mathcal{G}_1$ and $\mathcal{G}_2$ we can represent the sequence of genes in the target genome $\mathcal{G}_2$ as the identity string ι^n and map the genes in $\mathcal{G}_1$ accordingly in a string A.

Furthermore, if a continuous segment from $\mathcal{G}_1$ is not in $\mathcal{G}_2$ or conversely (i.e. there exists a continuous sequence of genes and intergenic regions between those genes that needs to be either removed or added), then this segment is modeled as only one element in the string, since an indel can insert or delete this whole segment at once. Since segments that are only in $\mathcal{G}_1$ are removed regardless of their content, we represent them by the character α and the sign is unnecessary. Except for α, we use integer numbers as the elements of the strings and, so, the set $\Sigma_A \cap \Sigma_{\iota^n}$ has only numbers. See an example in Fig. 1.

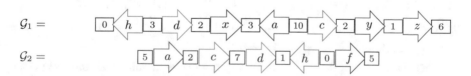

$$\mathcal{G}_1 =$$
$$\mathcal{G}_2 =$$

Fig. 1. Example of two genomes $\mathcal{G}_1$ and $\mathcal{G}_2$, where genes are represented by letters inside arrows, gene orientation is indicated by the orientation of the arrows, and the sizes of intergenic regions are represented by numbers inside squares. The genes of $\mathcal{G}_2$ are mapped as follows: a is mapped into $+1$; c is mapped into $+2$, d is mapped into $+3$, h is mapped into $+4$, and f is mapped into $+5$. So, the genome $\mathcal{G}_2$ is represented by $(\iota^n, \breve{\iota}^n)$, where $\iota^n = (+1 \ +2 \ +3 \ +4 \ +5)$, and $\breve{\iota}^n = (5, 2, 7, 1, 0, 5)$. The gene x and the segment that goes from y to z in $\mathcal{G}_1$ are not in $\mathcal{G}_2$ and, therefore, they are mapped into the character α. The genome $\mathcal{G}_1$ is represented by $(A, \breve{A})$, where $A = (+4 \ +3 \ \alpha \ −1 \ +2 \ \alpha)$, $\breve{A} = (0, 3, 2, 3, 10, 2, 6)$. The alphabets Σ_A and Σ_{ι^n} are equal to $\{1, 2, 3, 4, \alpha\}$ and $\{1, 2, 3, 4, 5\}$, respectively.

Now, let us define how rearrangements affect a genome $\mathcal{G} = (A, \breve{A})$, assuming that $|A| = n$. We denote by $\mathcal{G} \cdot \beta$ the resulting genome after applying a rearrangement β in $\mathcal{G}$. Similarly, we use $A \cdot \beta$ and $\breve{A} \cdot \beta$ to denote the effects of β on the string A and on the list of intergenic regions sizes $\breve{A}$, respectively.

A *reversal* $\rho_{(x,y)}^{(i,j)}$, with $1 \leq i \leq j \leq n$, $0 \leq x \leq \breve{A}_i$, and $0 \leq y \leq \breve{A}_{j+1}$, splits the intergenic regions $\breve{A}_i$ into (x, x') and $\breve{A}_{j+1}$ into (y, y'), where $x' = \breve{A}_i - x$ and $y' = \breve{A}_{j+1} - y$, and inverts the segment $(x'\ A_i\ \breve{A}_{i+1}\ A_{i+1}\ \ldots\ \breve{A}_j\ A_j\ y)$, changing the orientation of the genes in it. The genome $\mathcal{G} \cdot \rho_{(x,y)}^{(i,j)}$ is equal to $(A', \breve{A}')$ such that $A' = A \cdot \rho_{(x,y)}^{(i,j)} = (A_1\ \ldots\ A_{i-1} \underline{-A_j\ \ldots\ -A_i}\ A_{j+1} \ldots\ A_n)$ and $\breve{A}' = \breve{A} \cdot \rho_{(x,y)}^{(i,j)} = (\breve{A}_1, \ldots, \breve{A}_{i-1}, x + y, \underline{\breve{A}_j, \ldots, \breve{A}_{i+1}}, x' + y', \breve{A}_{j+2}, \ldots, \breve{A}_{n+1})$.

A reversal is a conservative rearrangement, that is, a rearrangement that does not alter the amount of genetic material in a genome. The non-conservative rearrangements considered in this model are the insertions and deletions.

An *insertion* is denoted by $\phi_x^{(i,S,\breve{S})}$, where (i) $0 \leq i \leq n$ and $0 \leq x \leq \breve{A}_{i+1}$, (ii) S is a sequence containing only characters belonging to $\Sigma_{\iota_n} \backslash \Sigma_A$, (iii) and $\breve{S}$ is a list of non-negative integers of size $|S| + 1$. The rearrangement $\phi_x^{(i,S,\breve{S})}$ inserts the sequence $(\breve{S}_1\ S_1\ \ldots\ \breve{S}_{|S|}\ S_{|S|}\ \breve{S}_{|S|+1})$, after the x-th nucleotide of the intergenic region $\breve{A}_{i+1}$, transforming $\mathcal{G}$ into $\mathcal{G} \cdot \phi_x^{(i,S,\breve{S})} = (A', \breve{A}')$ such that $A' = A \cdot \phi_x^{(i,S,\breve{S})} = (A_1\ \ldots\ A_i\ \underline{S_1\ \ldots\ S_{|S|}}\ A_{i+1}\ \ldots\ A_n)$ and $\breve{A}' = \breve{A} \cdot \phi_x^{(i,S,\breve{S})} = (\breve{A}_1, \ldots, \breve{A}_i, \underline{x + \breve{S}_1, \ldots, \breve{S}_{|S|}, \breve{S}_{|S|+1} + x'}, \breve{A}_{i+2}, \ldots, \breve{A}_{n+1})$, where $x' = \breve{A}_{i+1} - x$.

We note that if S is an empty sequence, the insertion only alters the intergenic region $\breve{A}_{i+1}$, changing it to $\breve{A}_{i+1} + \breve{S}_1$.

A deletion is denoted by $\psi_{(x,y)}^{(i,j)}$, such that $1 \leq i \leq j \leq n + 1$, $A_k = \alpha$ for $i \leq k < j$, $0 \leq x \leq \breve{A}_i$, and $0 \leq y \leq \breve{A}_j$. The rearrangement $\psi_{(x,y)}^{(i,j)}$ deletes the segment that starts after the x-th nucleotide of $\breve{A}_i$ and ends at the y-th nucleotide of $\breve{A}_j$, transforming $\mathcal{G}$ into $\mathcal{G} \cdot \psi_{(x,y)}^{(i,j)} = (A', \breve{A}')$ such that $A' = A \cdot \psi_{(x,y)}^{(i,j)} = (A_1\ \ldots\ A_{i-1}\ A_j\ \ldots\ A_n)$ and $\breve{A}' = \breve{A} \cdot \psi_{(x,y)}^{(i,j)} = (\breve{A}_1, \ldots, \breve{A}_{i-1}, x + y', \breve{A}_{j+1}, \ldots, \breve{A}_{n+1})$, where $y' = \breve{A}_j - y$.

When $i = j$, a deletion $\psi_{(x,y)}^{(i,j)}$ does not remove elements of A and only alters the intergenic region $\breve{A}_j$, so it must meet the constraint $0 \leq x \leq y \leq \breve{A}_j$.

The *reversal and indel distance* $d(\mathcal{G}_1, \mathcal{G}_2)$ is equal to the minimum number of reversals and indels that transform $\mathcal{G}_1$ into $\mathcal{G}_2$. In the distance problem studied in this work, given two genomes $\mathcal{G}_1 = (A, \breve{A})$ and $\mathcal{G}_2 = (\iota^n, \breve{\iota}^n)$, the goal is to find the value of $d(\mathcal{G}_1, \mathcal{G}_2)$.

3   Labeled Intergenic Breakpoint Graph

To represent an instance $\mathcal{G}_1 = (A, \breve{A})$ and $\mathcal{G}_2 = (\iota^n, \breve{\iota}^n)$ we adapted the *breakpoint graph* structure [2,9], adding information about both the intergenic regions sizes and the elements in $\Sigma_A \backslash \Sigma_{\iota^n}$ or $\Sigma_{\iota^n} \backslash \Sigma_A$. Oliveira *et al.* [10] also presented a structured based on the breakpoint graph, but it only deals with intergenic information and does not represent elements present in only one genome.

Given $\mathcal{G}_1 = (A, \breve{A})$ and $\mathcal{G}_2 = (\iota^n, \breve{\iota}^n)$, we extend the strings A and ι^n adding the elements $A_0 = +0$, $\iota_0^n = +0$, $A_{|A|+1} = +(n+1)$, and $\iota_{n+1}^n = +(n+1)$. The extra elements 0 and $n+1$ are not considered in the alphabet of the strings. In the next definitions, we assume that strings of an instance are in their extended form. Let π^A be a string containing the elements of A that also belong to ι^n preserving the same relative order they appear in A. For instance, if $A = (0\ 4\ \alpha\ 2\ 5\ \alpha\ 1\ 6)$ and $\iota^5 = (0\ 1\ \ldots\ 5\ 6)$, then $\pi^A = (0\ 4\ 2\ 5\ 1\ 6)$.

We define $next(x, \Sigma_A \cap \Sigma_{\iota^n})$, where $x \in (\Sigma_A \backslash \{\alpha\}) \cup \{0\}$, as follows:

$$next(x, \Sigma_A \cap \Sigma_{\iota^n}) = \begin{cases} min(y \in \Sigma_A \cap \Sigma_{\iota^n} | y > x), & \text{if } 0 \le x < max(\Sigma_A \cap \Sigma_{\iota^n}) \\ n+1, & \text{if } x = max(\Sigma_A \cap \Sigma_{\iota^n}) \end{cases}$$

In the following definitions, we consider that the size of π^A (not considering extended elements) is equal to m. The *labeled intergenic breakpoint graph* $G(\mathcal{G}_1, \mathcal{G}_2) = (V, E, w, \ell)$ is a graph where $V = \{+\pi_0^A, -\pi_1^A, +\pi_1^A, \ldots, -\pi_m^A, +\pi_m^A, -\pi_{m+1}^A\}$ is the set of vertices, E is the set of edges, $w : E \to \mathbb{Z}^*$ is a weight function that relates edges to intergenic regions sizes, and $\ell : E \to (\Sigma_{\iota^n} \backslash \Sigma_A) \cup \{\alpha\}$ is an edge labeling function.

The set of edges is divided into two categories: *origin edges* and *target edges*. The origin edges connect vertices using the position of the elements in A and the target edges connect vertices using the position of the elements in ι^n. For $1 \le i \le m+1$, there exists an origin edge $e_i = (+\pi_{i-1}^A, -\pi_i^A)$ whose weight $w(e)$ is equal to the sum of the intergenic regions between the elements π_{i-1}^A and π_i^A in $\mathcal{G}_1 = (A, \breve{A})$. The label of e_i is equal to $\ell(e_i) = \alpha$, if there exists at least one element α between π_{i-1}^A and π_i^A; and it is empty otherwise (i.e., $\ell(e_i) = \emptyset$). For every $x \in \Sigma_A \cup \{0\}$, there exists a target edge $e_x' = (+x, -next(x, \Sigma_A \cap \Sigma_{\iota^n}))$ whose weight $w(e_x')$ is equal to the sum of the intergenic regions between x and $next(x, \Sigma_A \cap \Sigma_{\iota^n})$ in $\mathcal{G}_2 = (\iota^n, \breve{\iota}^n)$. The label of e_x' is equal to $\ell(e_x') = x + 1$, if $x + 1 \neq next(x, \Sigma_A \cap \Sigma_{\iota^n})$, and it is empty otherwise.

Note that there exists $m+1$ origin edges and $m+1$ target edges in $G(\mathcal{G}_1, \mathcal{G}_2)$. Since each vertex of $G(\mathcal{G}_1, \mathcal{G}_2)$ is incident to one target edge and one origin edge, there exists a unique decomposition of $G(\mathcal{G}_1, \mathcal{G}_2)$ into cycles. Furthermore, for any cycle C of this decomposition, any pair of consecutive edges of C have different types (target or origin).

We draw the graph as shown in Fig. 2. The vertices are drawn from left to right following the sequence $+\pi_0^A, -\pi_1^A, +\pi_1^A, \ldots, -\pi_m^A, +\pi_m^A, -\pi_{m+1}^A$. Origin edges are represented as horizontal lines and target edges are represented as arcs. Labels and weights of edges appear above and below the edges, respectively. Also, labeled edges are drawn as dashed lines.

We say that an origin edge $e_i = (+\pi_{i-1}^A, -\pi_i^A)$ has index i and a target edge $e_x' = (+x, -next(x, \Sigma_A \cap \Sigma_{\iota^n}))$ has index x. Each cycle C is represented by the index of its edges $(o_1, t_1, o_2, t_2, \ldots, o_k, t_k)$ in the order they are traversed,

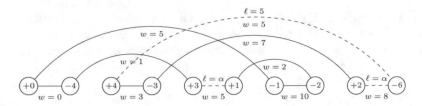

Fig. 2. Labeled intergenic breakpoint graph for $\mathcal{G}_1 = (A, \breve{A})$ and $\mathcal{G}_2 = (\iota^n, \breve{\iota}^n)$, where $A = (+0\ +4\ +3\ \alpha\ -1\ +2\ \alpha\ +6)$, $\breve{A} = (0, 3, 2, 3, 10, 2, 6)$, $n = 5$, $\iota^n = (+0\ +1\ +2\ +3\ +4\ +5\ +6)$, and $\breve{\iota}^n = (5, 2, 7, 1, 2, 3)$. We omit empty labels to simplify the graph drawing.

assuming that e_{o_1} is the rightmost origin edge and e_{o_1} is traversed from right to left. Furthermore, we say that a cycle with k origin edges is a k-cycle. A k-cycle is *trivial* if $k = 1$, *short* if $k = 2$, and *long* if $k \geq 3$.

For any cycle $C = (o_1, t_1, o_2, t_2, \ldots, o_k, t_k)$, we say that an origin edge is *convergent* if it is traversed from right to left, and we say that it is *divergent* if it is traversed from left to right.

A cycle C is *convergent* if all its origin edges are convergent; otherwise, C is *divergent*. Note that trivial cycles are convergent. Given a divergent cycle $C = (o_1, t_1, \ldots, o_k, t_k)$, we say that a pair of origin edges (e_{o_i}, e_{o_j}) is a *divergent pair* if one of these edges is convergent and the other is divergent.

We say that an edge e is *labeled*, if it has a non-empty label (i.e. $\ell(e) \neq \emptyset$), or *clean*, otherwise. Furthermore, we say that a cycle C is *labeled* if any of its edges is labeled. If all edges of a cycle C are clean, we say that C is *clean*.

We also classify cycles using the weight of its edges, which represent the intergenic regions sizes. A cycle $C = (o_1, t_1, \ldots, o_k, t_k)$ is *balanced* if $\sum_{i=1}^{k} w(e'_{t_i}) - \sum_{i=1}^{k} w(e_{o_i}) = 0$; otherwise, it is *unbalanced*. Furthermore, an unbalanced cycle is positive or negative. If $\sum_{i=1}^{k} w(e'_{t_i}) - \sum_{i=1}^{k} w(e_{o_i}) > 0$, then C is *positive*. If $\sum_{i=1}^{k} w(e'_{t_i}) - \sum_{i=1}^{k} w(e_{o_i}) < 0$, then C is *negative*.

Let $c(\mathcal{G}_1, \mathcal{G}_2)$ and $c^b_{clean}(\mathcal{G}_1, \mathcal{G}_2)$ be the number of cycles and the number of balanced and clean cycles in $G(\mathcal{G}_1, \mathcal{G}_2)$, respectively. Given a rearrangement β, $\Delta c(\mathcal{G}_1, \mathcal{G}_2, \beta) = (|\pi^A| + 1 - c(\mathcal{G}_1, \mathcal{G}_2)) - (|\pi^A \cdot \beta| + 1 - c(\mathcal{G}_1 \cdot \beta, \mathcal{G}_2))$, that is, $\Delta c(\mathcal{G}_1, \mathcal{G}_2, \beta)$ denotes the variation caused by the rearrangement β in the number of cycles relative to the number of origin edges in the graph. Given a rearrangement β, let $\Delta c^b_{clean}(\mathcal{G}_1, \mathcal{G}_2, \beta) = (|\pi^A| + 1 - c^b_{clean}(\mathcal{G}_1, \mathcal{G}_2)) - (|\pi^A \cdot \beta| + 1 - c^b_{clean}(\mathcal{G}_1 \cdot \beta, \mathcal{G}_2))$. We use analogous definitions for sequences of rearrangements, such as $\Delta c^b_{clean}(\mathcal{G}_1, \mathcal{G}_2, S)$ where S is a sequence of rearrangements.

Since $\mathcal{G}_1 = \mathcal{G}_2$ if, and only if, the graph $G(\mathcal{G}_1, \mathcal{G}_2)$ has only trivial cycles that are clean and balanced, the goal of our algorithm is to apply rearrangements that transform a labeled intergenic breakpoint graph into a graph containing only trivial cycles that are clean and balanced. In the following lemmas, we present how reversals and indels affect the number of cycles in the graph.

Lemma 1. *For any reversal ρ, we have $\Delta c(\mathcal{G}_1, \mathcal{G}_2, \rho) \in \{-1, 0, 1\}$ (Bafna and Pevzner [3]).*

Lemma 2. *For any reversal ρ, we have $\Delta c^b_{clean}(\mathcal{G}_1, \mathcal{G}_2, \rho) \leq 1$.*

Proof. From Lemma 1 we know that a reversal can increase the number of cycles by at most one, and we also know that the number of origin edges remains the same since no elements are added into the genome. In this scenario, the reversal acts on a cycle C breaking it into cycles C' and C'' [3]. If C is clean and balanced, in the best case, C' and C'' are also clean and balanced and $\Delta c^b_{clean}(\mathcal{G}_1, \mathcal{G}_2, \rho) = 1$. If C is labeled or unbalanced, then at least one of C' and C'' must be labeled or unbalanced. Therefore in the best case one of C' and C'' is clean and balanced, which results in $\Delta c^b_{clean}(\mathcal{G}_1, \mathcal{G}_2, \rho) = 1$. $\qquad\square$

Lemma 3. *For any indel β, $\Delta c(\mathcal{G}_1, \mathcal{G}_2, \beta) = 0$ and $\Delta c^b_{clean}(\mathcal{G}_1, \mathcal{G}_2, \beta) \leq 1$.*

Proof. Consider a deletion ψ. Since a deletion can only remove a segment containing elements in $\Sigma_A \setminus \Sigma_{\iota^n}$, a single origin edge is affected by ψ, but only the weight and label of this edge can change. The sets of vertices and edges of the graph remain the same. Consequently, the cycles of $G(\mathcal{G}_1 \cdot \rho, \mathcal{G}_2)$ remain the same which results in $\Delta c(\mathcal{G}_1, \mathcal{G}_2, \psi) = 0$. Let C be the cycle containing the affected edge. If this cycle is unbalanced or labeled, in the best scenario the deletion can turn this cycle into a balanced and clean cycle. Otherwise, the number of cycles that are balanced and clean remains the same. Therefore, $\Delta c^b_{clean}(\mathcal{G}_1, \mathcal{G}_2, \psi) \leq 1$.

Consider an insertion ϕ that inserts k elements in A. This insertion adds $2k$ vertices into the graph and replaces an origin edge of a cycle C by $k + 1$ origin edges, also adding k target edges into the graph. Since the new graph has $k + 1$ distinct origin edges in relation to $G(\mathcal{G}_1, \mathcal{G}_2)$ and at least one of these origin edges belongs to C, in the best scenario, at most k cycles are added into the graph. Therefore,

$$(|\pi^A| + 1 - c(\mathcal{G}_1, \mathcal{G}_2)) - (|\pi^A \cdot \phi| + 1 - c(\mathcal{G}_1 \cdot \phi, \mathcal{G}_2))$$
$$= (|\pi^A| - |\pi^A \cdot \phi|) - (c(\mathcal{G}_1, \mathcal{G}_2) - c(\mathcal{G}_1 \cdot \phi, \mathcal{G}_2)) = -k - (-k) = 0$$

In the best scenario, the cycle C becomes balanced and clean and all the k cycles added are also balanced and clean, which results in $\Delta c^b_{clean}(\mathcal{G}_1, \mathcal{G}_2, \phi) = 1$. Note that if any other cycle C' from $G(\mathcal{G}_1, \mathcal{G}_2)$ becomes balanced and clean after applying ϕ, then at least one of the new origin edges belongs to C' and the insertion could not have added k new cycles into the graph. Similarly, if x cycles different from C become balanced and clean, then at most $k - x$ cycles were added into the graph, which also results in $\Delta c^b_{clean}(\mathcal{G}_1, \mathcal{G}_2, \phi) \leq 1$. $\qquad\square$

The effect of the operations in the cycles of $G(\mathcal{G}_1, \mathcal{G}_2)$ gives the following lower bound for the distance of $\mathcal{G}_1$ and $\mathcal{G}_2$.

Lemma 4. *Given two genomes $\mathcal{G}_1 = (A, \breve{A})$ and $\mathcal{G}_2 = (\iota^n, \breve{\iota^n})$, we have that $d(\mathcal{G}_1, \mathcal{G}_2) \geq |\pi^A| + 1 - c^b_{clean}(\mathcal{G}_1, \mathcal{G}_2)$.*

Proof. Note that $\mathcal{G}_1 = \mathcal{G}_2$ if, and only if, the graph $G(\mathcal{G}_1, \mathcal{G}_2)$ has only trivial cycles that are balanced and clean (i.e., $|\pi^A| + 1 - c_{clean}^b(\mathcal{G}_1, \mathcal{G}_2) = 0$). Therefore, any sequence that transforms $\mathcal{G}_1$ into $\mathcal{G}_2$ must decrease $|\pi^A| + 1 - c_{clean}^b(\mathcal{G}_1, \mathcal{G}_2)$ to zero. By Lemmas 2 and 3, a rearrangement can decrease $|\pi^A| + 1 - c_{clean}^b(\mathcal{G}_1, \mathcal{G}_2)$ by at most one and, therefore, the bound follows. $\square$

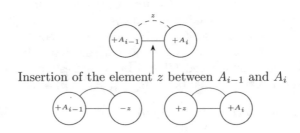

Insertion of the element z between A_{i-1} and A_i

Fig. 3. Insertion that transforms a labeled trivial cycle, which has a clean origin edge, into two clean trivial cycles, assuming that A_{i-1} is positive.

4   3-Approximation Algorithm

We present a 3-approximation algorithm for computing the distance by analyzing how we can increase the number of clean and balanced cycles in a labeled intergenic breakpoint graph.

Given two origin edges e_i and e_j, we say that a reversal applied to these two origin edges inverts the segment that goes from π_i^A to π_{j-1}^A in $\mathcal{G}_1$, and it may include α elements to the left of π_i^A or to the right of π_{j-1}^A.

Lemma 5. *If there exists a trivial cycle $C = (o_1, t_1)$, where $\ell(e_{o_1})$ is empty, such that (i) C is unbalanced and $\ell(e_{t_1}')$ is empty or (ii) C is non-negative and e_{t_1}' has a non-empty label, then there exists an indel β with $\Delta c_{clean}^b(\mathcal{G}_1, \mathcal{G}_2, \beta) = 1$.*

Proof. Since $\ell(e_{o_1})$ is empty, the edges e_{o_1} and e_{t_1}' connect vertices corresponding to adjacent elements in A. Let $+A_{i-1}$ and $-A_i$ be the vertices incident to these edges. We further divide our analysis into two cases.

(i) C is unbalanced and $\ell(e_{t_1}')$ is empty. If $w(e_{t_1}') > w(e_{o_1})$, then an insertion of $w(e_{t_1}') - w(e_{o_1})$ nucleotides in the intergenic region $\breve{A}_i$ turns C into a balanced and clean cycle. If $w(e_{t_1}') < w(e_{o_1})$, then a deletion of $w(e_{o_1}) - w(e_{t_1}')$ nucleotides in the intergenic region $\breve{A}_i$ turns C into a balanced and clean cycle. Since no elements are added to the string and C is turned into a balanced and clean cycle, there exists an indel β with $\Delta c_{clean}^b(\mathcal{G}_1, \mathcal{G}_2, \beta) = 1$.

(ii) C is non-negative and e_{t_1}' has a non-empty label. By our representation of the genomes, the label of e_{t_1}' is equal to $z = |A_{i-1}| + 1$. Let x be the intergenic region size between A_{i-1} and z in $\mathcal{G}_2$, and let y be the intergenic region size between z and A_i in $\mathcal{G}_2$. Note that A_{i-1} and A_i have the same sign since they form a trivial cycle. The insertion $\phi^{(i-1,S,\check{S})}_{\min(x,w(e_{o_1}))}$, where $S = (z)$, if A_{i-1} is positive, or $S = (-z)$, otherwise, turn C into two trivial cycles C' and C'', as shown in Fig. 3; and the intergenic list $\check{S} = (x',y')$, with $x' = x - \min(x, w(e_{o_1}))$ and $y' = y - (w(e_{o_1}) - \min(x, w(e_{o_1})))$, turn these two cycles into balanced and clean cycles. Since one element is added to the string and two balanced and clean cycles are created, there exists an indel β with $\Delta c^b_{clean}(\mathcal{G}_1, \mathcal{G}_2, \beta) = 1$. $\square$

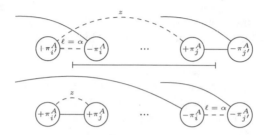

Fig. 4. Reversal applied to a divergent k-cycle C that transforms it into a trivial cycle C' and a $(k-1)$-cycle C''.

Lemma 6. *If there exists a trivial cycle C in $G(\mathcal{G}_1, \mathcal{G}_2)$ that is unbalanced or labeled, then there exists a sequence S_β with at most two rearrangements with $\Delta c^b_{clean}(\mathcal{G}_1, \mathcal{G}_2, S_\beta) = 1$.*

Proof. If the origin edge in C is labeled or if C is negative then a deletion ψ can remove the label of the origin edge and turn C into a non-negative cycle. The resulting trivial cycle is either (i) balanced and clean or (ii) it meets the conditions of Lemma 5. In the first case, we have $\Delta c^b_{clean}(\mathcal{G}_1, \mathcal{G}_2, \psi) = 1$. In the second case, there exists an indel β with $\Delta c^b_{clean}(\mathcal{G}_1, \mathcal{G}_2, \beta) = 1$. Otherwise, we know that the origin edge in C is clean and C is a non-negative cycle. Therefore, C satisfies Lemma 5 and there exists an indel β with $\Delta c^b_{clean}(\mathcal{G}_1, \mathcal{G}_2, \beta) = 1$. $\square$

Lemma 7. *If there exists a divergent labeled cycle in $G(\mathcal{G}_1, \mathcal{G}_2)$, then there exists a sequence S_β with at most two rearrangements with $\Delta c^b_{clean}(\mathcal{G}_1, \mathcal{G}_2, S_\beta) \geq 1$.*

Proof. Let $C = (o_1, t_1, \ldots, o_k, t_k)$ be a divergent k-cycle. Let $(e_{o_x}, e_{o_{x+1}})$ be a divergent pair such that x is minimum. A reversal ρ applied on these two origin edges $(e_{o_x}, e_{o_{x+1}})$ transforms C into a trivial cycle C' and a $(k-1)$-cycle C'', as shown in Fig. 4. Assume, without loss of generality, that the trivial cycle C' has the origin edge with index o_{x+1} in the new graph. If the edge $e_{o_{x+1}}$ is labeled, the reversal also includes the characters α in the origin edge $e_{o_{x+1}}$ to accumulate it into e_{o_x}. Also, if $w(e'_{t_x}) < w(e_{o_{x+1}})$, the reversal moves the exceeding weight from $e_{o_{x+1}}$ to e_{o_x} turning the trivial cycle C' into a balanced cycle; otherwise, the cycle C' is non-negative and the reversal does not move any weight from $e_{o_{x+1}}$. If C' is balanced and clean, then $\Delta c^b_{clean}(\mathcal{G}_1, \mathcal{G}_2, \rho) \geq 1$. Otherwise, C' meets the conditions of Lemma 5 and there exists an indel β with $\Delta c^b_{clean}(\mathcal{G}_1, \mathcal{G}_2, \beta) = 1$. $\square$

Lemma 8. *If there exists a clean divergent cycle C in $G(\mathcal{G}_1, \mathcal{G}_2)$, then there exists a rearrangement β with $\Delta c^b_{clean}(\mathcal{G}_1, \mathcal{G}_2, \beta) \geq 1$.*

Proof. If $C = (o_1, t_1, \ldots, o_k, t_k)$ is positive, then increasing the weight of any origin edge of C in $\sum_{i=1}^{k} w(e'_{t_i}) - \sum_{i=1}^{k} w(e_{o_i})$ units (i.e., an insertion ϕ of this amount in an intergenic region corresponding to the weight of an origin edge of C) turns C into a balanced cycle. In this way, there exists ϕ such that $\Delta c^b_{clean}(\mathcal{G}_1, \mathcal{G}_2, \phi) = 1$.

Otherwise, C is a negative or a balanced divergent cycle, and Oliveira et al. [10] proved that there exists a reversal ρ that splits C into C' and C'', increasing the number of balanced cycles in one unit. If C is negative, then either C' or C'' is balanced; otherwise, both C' and C'' are balanced. Since C is clean and the reversal is applied only in the origin edges of C, the cycles C' and C'' are also clean. Therefore, there exists a reversal ρ with $\Delta c^b_{clean}(\mathcal{G}_1, \mathcal{G}_2, \rho) = 1$. $\square$

To present the next lemma, we first define cycle intersection and open gates. Let e'_{t_1} be a target edge adjacent to origin edges with indexes x_1 and y_1, such that $x_1 < y_1$, and let e'_{t_2} be a target edge adjacent to origin edges with indexes x_2 and y_2, such that $x_2 < y_2$. We say that these target edges *intersect* if $x_1 < x_2 \leq y_1 < y_2$. Two cycles C and C' *intersect* if any target edge from C intersects with a target edge from C'.

A target edge e' from a non-trivial cycle C is an *open gate* if it does not intersect with any other target edge belonging to C. An open gate e' from C is *closed* if there exists another target edge that intersects with e', note that this target edge must be from another cycle from the definition of an open gate. All open gates in a breakpoint graph are closed [3].

Case 1: Cycle with no open gates.   Case 2: Cycle with open gates.

Fig. 5. How reversals can turn a convergent cycle into a divergent cycle.

Algorithm 1: 3-approximation algorithm for computing a minimum length sequence of rearrangements that transforms $\mathcal{G}_1$ into $\mathcal{G}_2$.

Input: Genomes $\mathcal{G}_1$ and $\mathcal{G}_2$
Output: A sequence of reversals and indels that transforms $\mathcal{G}_1$ into $\mathcal{G}_2$

1 Let $S_\beta \leftarrow \emptyset$
2 **while** $\mathcal{G}_1 \neq \mathcal{G}_2$ **do**
3   **if** $G(\mathcal{G}_1, \mathcal{G}_2)$ *has an unbalanced or labeled trivial cycle* $C = (o_1, t_1)$ **then**
4   **if** $\ell(e_{o_1})$ *is empty and* C *is non-negative* **then**
5   Let β be an indel according to Lemma 5
6   Apply β to $\mathcal{G}_1$ and append it to S_β
7   **else**
8   Let S_β' be a sequence of rearrangements according to Lemma 6
9   Apply S_β' to $\mathcal{G}_1$ and append it to S_β
10   **else if** $G(\mathcal{G}_1, \mathcal{G}_2)$ *has a labeled divergent cycle* **then**
11   Let S_β' be a sequence of rearrangements according to Lemma 7
12   Apply S_β' to $\mathcal{G}_1$ and append it to S_β
13   **else if** $G(\mathcal{G}_1, \mathcal{G}_2)$ *has a clean divergent cycle* **then**
14   Let S_β' be a sequence of rearrangements according to Lemma 8
15   Apply S_β' to $\mathcal{G}_1$ and append it to S_β
16   **else if** $G(\mathcal{G}_1, \mathcal{G}_2)$ *has a non-trivial convergent cycle* **then**
17   Let S_β' be a sequence of rearrangements according to Lemma 9
18   Apply S_β' to $\mathcal{G}_1$ and append it to S_β
19 **return** S_β

Lemma 9. *If there exists a convergent non-trivial cycle* $C = (o_1, t_1, \ldots, o_k, t_k)$ *and no divergent cycles in* $G(\mathcal{G}_1, \mathcal{G}_2)$, *then there exists a sequence* S_β *with at most three rearrangements such that* $\Delta c_{clean}^b(\mathcal{G}_1, \mathcal{G}_2, S_\beta) \geq 1$.

Proof. If C does not have open gates, then there exists distinct origin edges o_i, o_{i+1}, and o_j in C, with $1 \leq i < k$, such that $o_j \in [\min(o_i, o_{i+1})..\max(o_i, o_{i+1})]$. A reversal applied to o_i and o_j does not split C, since this is a convergent pair [9], and turns C into a divergent cycle (Fig. 5, Case 1).

Otherwise, there exists an open gate adjacent to origin edges o_i and o_{i+1}, which is closed by a target edge adjacent to o_x and o_{x+1} in a cycle C'. By the definition of open gates and intersection, either o_i or o_{i+1} is between o_x and o_{x+1}. Since there are no divergent cycles and no divergent pairs, a reversal applied to o_x and o_{x+1} does not split C' and turns C into a divergent cycle (Fig. 5, Case 2). Since C is now a divergent cycle, by Lemmas 7 and 8, there exists a sequence S_β with at most two rearrangements such that $\Delta c_{clean}^b(\mathcal{G}_1, \mathcal{G}_2, S_\beta) \geq 1$. $\qquad\square$

Using the results from Lemmas 5 to 9, we present Algorithm 1 that has an approximation factor of 3, as shown in Theorem 10.

Theorem 10. *Algorithm 1 is a 3-approximation algorithm for the Reversal and Indel Distance Considering Gene Order and Intergenic Regions problem.*

Proof. Given two genomes $\mathcal{G}_1 = (A, \breve{A})$ and $\mathcal{G}_2 = (\iota^n, \breve{\iota}^n)$, at each iteration, the algorithm uses up to 3 operations to decrease the value of $|\pi^A| + 1 - c^b_{clean}(\mathcal{G}_1, \mathcal{G}_2)$. Therefore, the sequence of rearrangements returned by the algorithm has length less than or equal to $3(|\pi^A| + 1 - c^b_{clean}(\mathcal{G}_1, \mathcal{G}_2))$. In this way, by Lemma 4, this algorithm has the approximation factor of 3.   □

5   Conclusions

We introduced a new rearrangement distance variation that extends and combines previous studies in the literature, where the compared genomes can have a distinct set of genes and using both gene order and intergenic region sizes on the representation. We presented the labeled intergenic breakpoint graph structure, which is based on the well-known breakpoint graph, to provide lower bounds for the distance and to develop a 3-approximation algorithm.

For future work, it is possible to extend the rearrangement model considering other rearrangements, such as transpositions, and study the complexity of these models.

Acknowledgments. This work was supported by the Coordenação de Aperfeiçoamento de Pessoal de Nível Superior - Brasil (CAPES) - Finance Code 001, the National Council of Technological and Scientific Development, CNPq (grants 140272/2020-8 and 425340/2016-3), and the São Paulo Research Foundation, FAPESP (grants 2013/08293-7, 2015/11937-9, 2017/12646-3, and 2019/27331-3).

References

1. Alexandrino, A.O., Oliveira, A.R., Dias, U., Dias, Z.: Genome rearrangement distance with reversals, transpositions, and indels. J. Comput. Biol. **28**(3), 235–247 (2021)
2. Bafna, V., Pevzner, P.A.: Sorting permutations by transpositions. In: Proceedings of the Sixth Annual ACM-SIAM Symposium on Discrete Algorithms (SODA 1995), pp. 614–623. Society for Industrial and Applied Mathematics, Philadelphia (1995)
3. Bafna, V., Pevzner, P.A.: Genome rearrangements and sorting by reversals. SIAM J. Comput. **25**(2), 272–289 (1996)
4. Biller, P., Guéguen, L., Knibbe, C., Tannier, E.: Breaking good: accounting for fragility of genomic regions in rearrangement distance estimation. Genome Biol. Evol. **8**(5), 1427–1439 (2016)
5. Biller, P., Knibbe, C., Beslon, G., Tannier, E.: Comparative genomics on artificial life. In: Beckmann, A., Bienvenu, L., Jonoska, N. (eds.) CiE 2016. LNCS, vol. 9709, pp. 35–44. Springer, Cham (2016). https://doi.org/10.1007/978-3-319-40189-8_4
6. Brito, K.L., Jean, G., Fertin, G., Oliveira, A.R., Dias, U., Dias, Z.: Sorting by genome rearrangements on both gene order and intergenic sizes. J. Comput. Biol. **27**(2), 156–174 (2020)
7. Caprara, A.: Sorting permutations by reversals and Eulerian cycle decompositions. SIAM J. Discrete Math. **12**(1), 91–110 (1999)
8. Fertin, G., Labarre, A., Rusu, I., Tannier, É., Vialette, S.: Combinatorics of Genome Rearrangements. Computational Molecular Biology. MIT Press, London (2009)

9. Hannenhalli, S., Pevzner, P.A.: Transforming cabbage into turnip: polynomial algorithm for sorting signed permutations by reversals. J. ACM **46**(1), 1–27 (1999)
10. Oliveira, A.R., et al.: Sorting Signed Permutations by Intergenic Reversals. IEEE/ACM Trans. Comput. Biol. Bioinform. (2020)
11. Willing, E., Stoye, J., Braga, M.D.: Computing the inversion-indel distance. IEEE/ACM Trans. Comput. Biol. Bioinform. (2020)

Reversals Distance Considering Flexible Intergenic Regions Sizes

Klairton Lima Brito[1]([⊠])[iD], Alexsandro Oliveira Alexandrino[1][iD],
Andre Rodrigues Oliveira[1][iD], Ulisses Dias[2][iD], and Zanoni Dias[1][iD]

[1] Institute of Computing, University of Campinas, Campinas, Brazil
{klairton,alexsandro,andrero,zanoni}@ic.unicamp.br
[2] School of Technology, University of Campinas, Limeira, Brazil
ulisses@ft.unicamp.br

Abstract. Most mathematical models for genome rearrangement problems have considered only gene order. In this way, the rearrangement distance considering some set of events, such as reversal events, is commonly defined as the minimum number of rearrangement events that transform the gene order from a genome $\mathcal{G}_1$ into the gene order from a genome $\mathcal{G}_2$. Recent works initiate incorporating more information such as the sizes of the intergenic regions (i.e., number of nucleotides between pairs of consecutive genes), which yields good results for estimated distances on realistic data. In these models, besides transforming the gene order, the sequence of rearrangement events must transform the list of intergenic regions sizes from $\mathcal{G}_1$ into the list of intergenic regions sizes from $\mathcal{G}_2$ (target list). We study a new variation where each value from the target list is flexible, in the sense that each target intergenic region size is a range of acceptable values. We consider the well-known reversals distance and present a 2-approximation algorithm, alongside NP-hardness proof. Our results rely on the Flexible Weighted Cycle Graph, adapted from the breakpoint graph to deal with flexible intergenic regions sizes.

Keywords: Genome rearrangements · Intergenic regions · Reversals distance

1 Introduction

Seminal works on genome rearrangement distance have used only gene order information to model genomes and assumed no repeated genes. In this way, each genome is modeled as a permutation of integers, where each element represents a gene and the orientation of a gene is represented by a plus or minus sign.

When using signed permutations, the problem of transforming one genome into another using rearrangements is equivalent to the problem of sorting a permutation by rearrangements. Most works in this area assume a parsimonious scenario and seek for minimum-length sequences. One of the most important of these rearrangement problems is the Sorting Signed Permutations by Reversals.

© Springer Nature Switzerland AG 2021
C. Martín-Vide et al. (Eds.): AlCoB 2021, LNBI 12715, pp. 134–145, 2021.
https://doi.org/10.1007/978-3-030-74432-8_10

A *reversal* is a rearrangement event that inverts a segment from the genome, flipping the orientation of the genes in the affected segment. Hannenhalli and Pevzner [9] presented a polynomial algorithm for this problem and, later, other studies improved the time complexity of this algorithm [1].

Recently, in addition to gene order, studies argued that incorporating intergenic regions sizes (i.e., number of nucleotides between pairs of consecutive genes) in the mathematical model of genomes gives better distance estimators in realistic data [4,5].

After that, studies began to consider problems in which a genome is modeled as a permutation, representing gene order, alongside a list of intergenic regions sizes. Now, the rearrangement distance problems seek for a minimum-length sequence of rearrangements that transform both the permutation and intergenic list from G_1 into the permutation and intergenic list from G_2. Some of these problems are (i) Reversals Distance, which is NP-hard on both signed and unsigned permutations [6,10]. The best results are a 2-approximation [10] and a 4-approximation [6] for the signed and unsigned case, respectively; (ii) Transpositions Distance, which is NP-hard and has a 3.5-approximation algorithm [11]. A transposition is a rearrangement event that changes the position of two adjacent segments of a genome; (iii) Double Cut and Join (DCJ) Distance, which is NP-hard and has a 4/3-approximation algorithm [7]. A DCJ is a generic event that splits a genome in two positions creating four endpoints and joins them in any way.

In this work, we investigate the reversals distance considering that each element in the target intergenic list is a range of acceptable values. This problem is named SORTING BY REVERSALS WITH FLEXIBLE INTERGENIC REGIONS (FSBR) and we deal with signed permutations (gene orientation is known). Our main results are a 2-approximation algorithm and a lower bound for the distance, which were achieved by using an adaption of the well-known breakpoint graph [9] called Flexible Weight Cycle Graph. We also present an NP-hardness proof using a reduction from the strict case (each element in the target intergenic list is a single value), which is called SORTING BY REVERSALS WITH INTERGENIC REGIONS (SBR).

This work is organized as follows. In Sect. 2, we introduce the basic notation and definitions used in the problems and the algorithm. In Sect. 3, we present a lower bound for this problem and a 2-approximation algorithm. At last, in Sect. 4, we conclude the paper and give directions for future work.

2   Definitions

In this section, we introduce definitions and concepts used throughout the paper. We will assume a **source genome** and a **target genome** sharing the same set of genes, with n unique genes, and $n + 1$ intergenic regions (i.e., nucleotides between each pair of genes and in both extremities of the genome).

We represent the information of the source and target genomes using two structures. First, we map the sequence of genes (both order and orientation) from the target genome into a specific permutation called **identity permutation**, denoted by $\iota = (+1 \; +2 \; \ldots \; +n)$, and each element ι_i of the permutation receives a "+" sign that indicates the orientation of the gene at position i. We then use a permutation π to reflect the information about gene order and orientation from the source genome, using labels according to ι. An element π_i has a "+" sign if the gene represented by π_i has the same orientation in both genomes and, otherwise, it has a "$-$" sign.

Genome rearrangement events that break genes often debilitate the individual, while events that affect intergenic regions do not. For this reason, we extract the information about the number of nucleotides in each intergenic region, which we call **size**, and represent the intergenic regions from the source genome as a list of non-negative integer numbers $\breve{\pi} = (\breve{\pi}_1, \ldots, \breve{\pi}_{n+1})$, such that $\breve{\pi}_i$ indicates the size of the intergenic region in the left of the i-th gene from π, if $i \leq n$, or after the last gene otherwise, if $i = n + 1$. Thus, we represent the source genome as $(\pi, \breve{\pi})$.

To make the model more flexible regarding the constraints about sizes of intergenic regions in the target genome, we map the size of intergenic regions using a range of possible values. In this way, we use two lists of non-negative integer numbers $\iota^{\min}$ and $\iota^{\max}$, such that $\iota_i^{\min} \leq \iota_i^{\max}$ for $1 \leq i \leq n + 1$, which indicate the minimum and maximum sizes allowed for each intergenic region, respectively. Therefore, the target genome is represented by $(\iota, \iota^{\min}, \iota^{\max})$. Since the identity permutation ι can be obtained given the information about the size of the permutation π, an instance I consists of four elements $(\pi, \breve{\pi}, \iota^{\min}, \iota^{\max})$. In the following definition, we show how an intergenic reversal affects a genome.

Definition 1. *Given a genome* $(\pi, \breve{\pi})$, *an **intergenic reversal** $\rho_{(x,y)}^{(i,j)}$, such that $1 \leq i \leq j \leq n$, $\{x, y\} \subset \mathbb{N}_0$, $x \leq \breve{\pi}_i$, and $y \leq \breve{\pi}_{j+1}$, splits the intergenic regions $\breve{\pi}_i$ (into x and x', with $x' = \breve{\pi}_i - x$) and $\breve{\pi}_{j+1}$ (into y and y', with $y' = \breve{\pi}_{j+1} - y$), and inverts the segment $(x', \pi_i, \breve{\pi}_{i+1}, \ldots, \breve{\pi}_j, \pi_j, y)$ swapping the sign of the elements π_s for $i \leq s \leq j$. Formally, an intergenic reversal $\rho_{(x,y)}^{(i,j)}$ applied into a genome $(\pi, \breve{\pi})$ results in the genome $(\pi', \breve{\pi}')$, such that $\breve{\pi}_i' = x + y$ and $\breve{\pi}_{j+1}' = x' + y'$:*

$$(\pi, \breve{\pi}) = (\breve{\pi}_1, +\pi_1 \ldots, +\pi_{i-1}, \underline{\breve{\pi}_i, +\pi_i, \breve{\pi}_{i+1} \ldots, \breve{\pi}_j, +\pi_j}, \breve{\pi}_{j+1}, +\pi_{j+1}, \ldots, +\pi_n, \breve{\pi}_{n+1})$$

$$(\pi', \breve{\pi}') = (\breve{\pi}_1, +\pi_1 \ldots, +\pi_{i-1}, \underline{\breve{\pi}_i', -\pi_j, \breve{\pi}_j \ldots, \breve{\pi}_{i+1}, -\pi_i, \breve{\pi}_{j+1}'}, +\pi_{j+1}, \ldots, +\pi_n, \breve{\pi}_{n+1})$$

From now on, we will refer to an intergenic reversal simply as a reversal. Note that a reversal is a **conservative event**, so it does not insert or remove nucleotides. Thus, we consider that an instance $(\pi, \breve{\pi}, \iota^{\min}, \iota^{\max})$ is **valid** if the following inequality is satisfied:

$$\sum_{\iota_i^{\min} \, \in \, \iota^{\min}} \iota_i^{\min} \leq \sum_{\breve{\pi}_i \, \in \, \breve{\pi}} \breve{\pi}_i \leq \sum_{\iota_i^{\max} \, \in \, \iota^{\max}} \iota_i^{\max}$$

Our goal is to transform a source genome $(\pi, \breve{\pi})$ into a target genome $(\iota, \iota^{\min}, \iota^{\max})$ attending the constraints regarding the minimum and maximum sizes allowed for each intergenic region. Note that $\iota^{\min}$ and $\iota^{\max}$ are not affected by any rearrangement event. In the following, we formally describe the problem addressed in this work.

Sorting by Reversals with Flexible Intergenic Regions (FSbR)

Input: A permutation π and three lists $\breve{\pi}$, $\breve{\iota}^{\min}$, and $\breve{\iota}^{\max}$.

Task: Find the shortest sequence Q of intergenic reversals that turns π into ι and $\breve{\pi}$ into $\breve{\pi}'$ such that $\breve{\iota}_i^{\min} \leq \breve{\pi}_i' \leq \breve{\iota}_i^{\max}$ for all $\breve{\pi}_i' \in \breve{\pi}'$.

The minimum number of reversals capable of transforming a source genome $(\pi, \breve{\pi})$ into a target genome $(\iota, \breve{\iota}^{\min}, \breve{\iota}^{\max})$ is called **distance**, and it is denoted by $d_{\text{FSbR}}(\pi, \breve{\pi}, \breve{\iota}^{\min}, \breve{\iota}^{\max})$.

The **extended form** of a permutation π includes two new elements in the extremities, $\pi_0 = 0$ and $\pi_{n+1} = +(n+1)$. These new elements are not affected by rearrangement events. From now on, we assume that any permutation is in its extended form, and we will refer to it simply as permutation.

2.1   Flexible Weighted Cycle Graph

To handle the information of an instance $(\pi, \breve{\pi}, \breve{\iota}^{\min}, \breve{\iota}^{\max})$ in a single structure, we adapted the **breakpoint graph** [3,8] and called it **flexible weighted cycle graph**. This structure allows us to derive bounds and develop algorithms for the FSbR problem.

Given an instance $I = (\pi, \breve{\pi}, \breve{\iota}^{\min}, \breve{\iota}^{\max})$, $\mathcal{G}(I) = (V, E, w_b, \ell_b, w_g^{\min}, w_g^{\max})$ is a graph in which $V = \{+\pi_0, -\pi_1, +\pi_1, -\pi_2, +\pi_2, \ldots, -\pi_n, +\pi_n, -\pi_{n+1}\}$ and E represent the set of vertices and edges, respectively. The set $E = E_b \cup E_g$ contains two types of edges, **black** and **gray**. Furthermore, $w_b : E_b \to \mathbb{N}_0$ is a weight function that maps black edges to values corresponding with the list $\breve{\pi}$, and $\ell_b : E_b \to \{1, \ldots, (n+1)\}$ is a labeling function. The weighted functions $w_g^{\min} : E_g \to \mathbb{N}_0$ and $w_g^{\max} : E_g \to \mathbb{N}_0$ link gray edges to the minimum and maximum size allowed in the intergenic regions of the target genome, respectively.

The sets of black and gray edges are defined as follows. The black edge set E_b is $\{(-\pi_i, +\pi_{i-1}) : 1 \leq i \leq n + 1\}$, such that $\ell_b((-\pi_i, +\pi_{i-1})) = i$ and $w_b((-\pi_i, +\pi_{i-1})) = \breve{\pi}_i$, for $1 \leq i \leq n+1$. The gray edge set E_g is $\{(+(i-1), -i) : 1 \leq i \leq n + 1\}$, such that $w_g^{\min}((+(i-1), -i)) = \breve{\iota}_i^{\min}$ and $w_g^{\max}((+(i-1), -i)) = \breve{\iota}_i^{\max}$, for $1 \leq i \leq n + 1$. Note that black and gray edges represent the genes adjacencies in the source and target genomes, respectively. Besides, note that the graph $\mathcal{G}(I)$ is composed of $2n + 2$ edges, being $n + 1$ black and $n + 1$ gray. Each vertex has degree two by the incidence of one black and one gray edge. Thus, there is a unique decomposition of $\mathcal{G}(I)$ in cycles with alternating edge colors.

There are different ways to draw the graph $\mathcal{G}(I)$. However, for convenience and to identify the cycles uniquely, we adopted the following standard method. We place the vertices in a horizontal line according to the permutation π. We draw the black edges horizontally and the grey edges as arcs over the vertices.

Each cycle C is represented as a list of black edge labels $(c^1, c^2, \ldots, c^k)$ in the order they are traversed. To make this representation unique, we assume that

c^1 is the label with the highest value in the cycle, and that it is traversed from right to left. Cycles with one, two, and three or more black edges are named as trivial, short, and long, respectively.

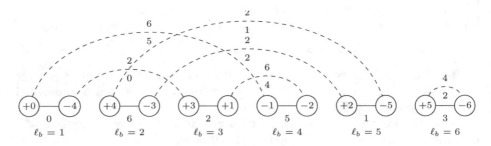

Fig. 1. An example of a flexible weighted cycle graph created from the instance $I = (\pi, \breve{\pi}, \iota^{\min}, \iota^{\max})$ where $\pi = (+4\ +3\ -1\ +2\ +5)$, $\breve{\pi} = (0, 6, 2, 5, 1, 3)$, $\iota^{\min} = (5, 4, 2, 0, 1, 2)$, and $\iota^{\max} = (6, 6, 2, 2, 2, 4)$. Circles represent vertices, while continuous and dashed lines indicate black and gray edges, respectively. Numbers below and above each gray edge represent minimum and maximum size allowed, respectively. Numbers below black edges indicate the weight associated with each of them. The graph $\mathcal{G}(I)$ has three cycles, $C_1 = (4, 1, 3)$, $C_2 = (5, 2)$, and $C_3 = (6)$. The cycles C_1 and C_2 are unstable, while C_3 is stable ($\mathcal{S} = \{C_3\}$ and $\mathcal{U} = \{C_1, C_2\}$).

Given a cycle C, $E_b(C)$ and $E_g(C)$ denote the set of black and gray edges that belong to the cycle C, respectively. We denote by $W_b(C) = \sum_{e \in E_b(C)} w_b(e)$, $W_g^{\min}(C) = \sum_{e \in E_g(C)} w_g^{\min}(e)$, and $W_g^{\max}(C) = \sum_{e \in E_g(C)} w_g^{\max}(e)$ the total weight (i.e., the sum of the weights of the black edges), total minimum weight (i.e., the sum of the minimum weights of the gray edges), and total maximum weight (i.e., the sum of the maximum weights of the gray edges) of the cycle C, respectively.

A cycle C is called **stable** if $W_g^{\min}(C) \leq W_b(C) \leq W_g^{\max}(C)$, and **unstable** otherwise. In other words, a stable cycle indicates that the total weight is enough to satisfy the constraints regarding the total minimum and maximum weights imposed by their gray edges. We define the sets of stable and unstable cycles in $\mathcal{G}(I)$ as $\mathcal{S}$ and $\mathcal{U}$, respectively. Figure 1 shows an example of the flexible weighted cycle graph.

Observe that to reach the target genome, it is necessary to remove all the unstable cycles from the graph and obtain $n+1$ stable cycles. Depending on the distribution of the nucleotides and minimum and maximum size constraints in the gray edges, some of the stable cycles must be affected as well. In fact, there are two cases where this occurs:

$$\text{(i)}\ \sum_{C \in \mathcal{U}} W_b(C) < \sum_{C \in \mathcal{U}} W_g^{\min}(C)$$

$$\text{(ii)}\ \sum_{C \in \mathcal{U}} W_b(C) > \sum_{C \in \mathcal{U}} W_g^{\max}(C)$$

In case (i), the total number of nucleotides in the black edges is insufficient to fill the minimum size required by the gray edges constraints. In case (ii), we have the opposite, the total number of nucleotides in the black edges exceeds the maximum size required by the gray edges constraints. Note that if the set $\mathcal{U}$ is empty, then none of the cases will occur. However, if the set $\mathcal{U}$ is not empty, only one of these cases may occur. In the following, we present definitions considering that either (i) or (ii) is true.

Definition 2. *A stable cycle is **auxiliary** if it must receive or transfer nucleotides to another cycle, and it is **free** otherwise.*

Definition 3. *Given a cycle C, we define $gap_{\min}(C) = W_b(C) - W_g^{\min}(C)$ as the number of the nucleotides exceeding the total minimum weight of the cycle C. Similarly, we define $gap_{\max}(C) = W_g^{\max}(C) - W_b(C)$ as the number of the nucleotides allowed to be inserted in C and not exceed the total maximum weight of C.*

Note that the number of auxiliary cycles depends on which case the instance fits in. If it is case (i), then a minimal set of auxiliary cycles $\mathcal{A}$ can be composed of the minimum number of cycles such that the following constraint is fulfilled:

$$\sum_{C \in \mathcal{A}} gap_{\min}(C) + \sum_{C \in \mathcal{U}} gap_{\min}(C) \geq 0$$

If it is case (ii), then a minimal set of auxiliary cycles $\mathcal{A}$ can be composed of the minimum number of cycles such that the following constraint is fulfilled:

$$\sum_{C \in \mathcal{A}} gap_{\max}(C) + \sum_{C \in \mathcal{U}} gap_{\max}(C) \geq 0$$

Note that in both cases, the set $\mathcal{A}$ can be easily obtained after sorting the stable cycles by the $gap_{\min}$ and $gap_{\max}$ values, in a decreasing way, considering the cases (i) and (ii), respectively. Then, following the decreasing order, the cycles are labeled as auxiliary until they satisfy the constraint. The set of free cycles $\mathcal{F}$ is obtained by $\mathcal{S} - \mathcal{A}$.

Given an instance $I = (\pi, \breve{\pi}, \breve{\iota}^{\min}, \breve{\iota}^{\max})$, we denote as $f(\mathcal{G}(I)) = |\mathcal{F}| = |\mathcal{S}| - |\mathcal{A}|$ the total number of free cycles in $\mathcal{G}(I)$. The variation in the number of free cycles after applying a rearrangement operation γ is denoted by $\Delta f(\mathcal{G}(I), \gamma)$ and defined as $\Delta f(\mathcal{G}(I), \gamma) = f(\mathcal{G}(I')) - f(\mathcal{G}(I))$, such that the instance I' is obtained after applying γ in the source genome $(\pi, \breve{\pi})$.

Remark 4. An instance $I = (\pi, \breve{\pi}, \breve{\iota}^{\min}, \breve{\iota}^{\max})$ such that $f(\mathcal{G}(I)) = n + 1$ implies that $\pi = \iota$ and $\breve{\iota}_i^{\min} \leq \breve{\pi}_i \leq \breve{\iota}_i^{\max}$ for all $\breve{\pi}_i \in \breve{\pi}$.

Figure 2 shows an example of a flexible weighted cycle graph created from the instance $I = (\pi = (+3 \ +2 \ +1 \ +4 \ +5), \breve{\pi} = (1, 2, 0, 2, 6, 2), \breve{\iota}^{\min} = (3, 2, 4, 0, 2, 1), \breve{\iota}^{\max} = (4, 3, 5, 4, 6, 2))$.

Observe that Fig. 2 shows the case (i): $1 = W_b(C_1) < W_g^{\min}(C_1) = 7$, where the only unstable cycle (C_1) needs to receive nucleotides to be turned into stable. Note that $gap_{\min}(C_2) = 2$, $gap_{\min}(C_3) = 4$, and $gap_{\min}(C_4) = 1$. Therefore, we have $\mathcal{A} = \{C_2, C_3\}$ and $\mathcal{F} = \{C_4\}$.

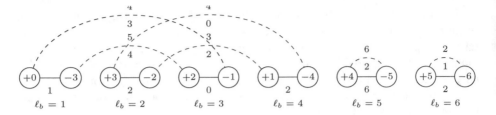

Fig. 2. An exemple of a flexible weighted cycle graph with four cycles $C_1 = (3, 1)$, $C_2 = (4, 2)$, $C_3 = (5)$, and $C_4 = (6)$. In this example, we have $\mathcal{U} = \{C_1\}$ and $\mathcal{S} = \{C_2, C_3, C_4\}$.

3   Results

In this section, we show a lower bound and describe how to obtain an algorithm with an approximation factor of 2 for the FSBR problem. Initially, we have to describe the SORTING BY REVERSALS WITH INTERGENIC REGIONS (SBR) problem, which is a variation of the FSBR problem with strict constraints regarding the size of the intergenic regions in the target genome. We developed an approximation algorithm for the FSBR problem based on an algorithm for the SBR problem. In the following, we describe formally the SBR problem.

SORTING BY REVERSALS WITH INTERGENIC REGIONS (SBR)
Input: A permutation π and two lists $\breve{\pi}$ and $\breve{\iota}$ such that $\sum_{\breve{\pi}_i \in \breve{\pi}} \breve{\pi}_i = \sum_{\breve{\iota}_i \in \breve{\iota}} \breve{\iota}_i$.
Task: Find the shortest sequence Q of intergenic reversals that turns π into ι and $\breve{\pi}$ into $\breve{\pi}'$ such that $\breve{\pi}'_i = \breve{\iota}_i$ for all $\breve{\pi}'_i \in \breve{\pi}'$.

Observe that the SBR problem has a strict constraint regarding the size of the intergenic regions, such that we have to transform π in ι and obtain a list of intergenic regions sizes $\breve{\pi}'$ that equals to $\breve{\iota}$. Oliveira *et al.* [10] showed that the SBR problem is NP-hard. Note that the FSBR problem generalizes the SBR problem, since we can reduce any instance of the SBR problem by keeping the same source genome and modifying only the size of the intergenic regions of the target genome to $\breve{\iota}^{\min} = \breve{\iota}^{\max} = \breve{\iota}$. Thus, we obtain the following lemma.

Lemma 5. *The FSBR problem is NP-hard.*

Now, we describe the structure called **weighted cycle graph**, which is similar to the flexible weighted cycle graph structure, used to obtain an algorithm for the SBR problem. Given an instance $I = (\pi, \breve{\pi}, \breve{\iota})$ of the SBR problem, $\mathcal{G}(I) = (V, E, w, l_b)$ is a graph in which $V = \{+\pi_0, -\pi_1, +\pi_1, -\pi_2, +\pi_2, \ldots, -\pi_n, +\pi_n, -\pi_{n+1}\}$ and E represent the set of vertices and edges, respectively. The set of edges $E = E_b \cup E_g$ is divided into black and gray edges. The weighted function $w : E \rightarrow \mathbb{N}_0$ associates the sizes of the intergenic regions of the source and target genome with weights in the edges. The labeling function

$\ell_b : E_b \rightarrow \{1, \ldots, (n+1)\}$ associates labels to the black edges. There is a black edge $(-\pi_i, +\pi_{i-1})$, for $1 \le i \le n+1$, with $w((-\pi_i, +\pi_{i-1})) = \breve{\pi}_i$ and $\ell_b((-\pi_i, +\pi_{i-1})) = i$. There is a gray edge $(+(i-1), -i)$, for $1 \le i \le n+1$, with $w((+(i-1), -i)) = \breve{\iota}_i$. For simplicity and unique identification, the disposal of the elements (vertices and edges) and the cycle representation follow the same standard procedure explained in Sect. 2.1.

Given a cycle C, the sets of black and gray edges of C are denoted by $E_b(C)$ and $E_g(C)$, respectively. A cycle C is **balanced** if $\sum_{e \in E_b(C)} w(e) = \sum_{e' \in E_g(C)} w(e')$ and **unbalanced** otherwise. In other words, a cycle is balanced if the sum of the weights of its black edges and the sum of the weights of its gray edges are equal. The number of balanced cycles in $\mathcal{G}(I)$ is denoted by $b(\mathcal{G}(I))$. Figure 3 shows an example of the weighted cycle graph created from the instance $I = (\pi = (+4 \ +3 \ -1 \ +2 \ +5), \breve{\pi} = (3, 1, 5, 0, 2, 3), \breve{\iota} = (5, 4, 2, 0, 1, 2))$. The graph $\mathcal{G}(I)$ has three cycles, $C_1 = (4, 1, 3)$, $C_2 = (5, 2)$, and $C_3 = (6)$. The cycles C_1 and C_3 are unbalanced while C_2 is balanced. The number of balanced cycles $b(\mathcal{G}(I))$ is one.

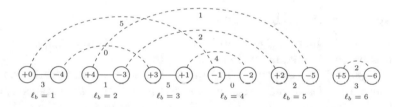

Fig. 3. An example of the weighted cycle graph, the black and gray edges are represented by solid and dashed lines, respectively.

Considering an instance $I = (\pi, \breve{\pi}, \breve{\iota})$ of the SBR problem, the minimum number of reversals capable to transform a source genome into a target genome is denoted by $d_{\text{SBR}}(I)$.

Proposition 6 (Theorem 3 [10]). *Given an instance $I = (\pi, \breve{\pi}, \breve{\iota})$ of the SBR problem, we have $d_{\text{SBR}}(I) \ge (n+1) - b(\mathcal{G}(I))$.*

3.1   Lower Bound

In this section, we obtain a lower bound for the FSBR problem based on the flexible weighted cycle graph structure.

Lemma 7. $\Delta f(\mathcal{G}(\pi, \breve{\pi}, \breve{\iota}^{\min}, \breve{\iota}^{\max}, \rho)) \le 1$ *for any reversal ρ.*

Proof. Since any reversal affects two black edges, there are only two possibilities: the black edges affected belong to (i) the same cycle or (ii) different cycles.

If the black edges belong to the same cycle, then C will be split into two new cycles C_1 and C_2, in the best case. The cycle C can be unstable, auxiliary, or free. If C is auxiliary or unstable, necessarily C_1 or C_2 will be auxiliary or unstable as well, and the number of free cycles increases by one. If C is free, the best case will generate C_1 and C_2 as free cycles, and the number of free cycles increases by one.

If the black edges belong to different cycles C_1 and C_2, then these two cycles will form a new cycle C_3 [2]. The cases where the number of free cycles can be increased occur when both C_1 and C_2 are unstable or when C_1 and C_2 are unstable and auxiliary, no matter the order. Note that, in both cases, in the best scenario C_3 is a free cycle.

In both cases, the number of free cycles increases at most by one and the lemma follows.   □

Lemma 8. *Given an instance* $I = (\pi, \breve{\pi}, \breve{\iota}^{\min}, \breve{\iota}^{\max})$ *of the* FSBR *problem, we have* $d_{\mathrm{FSBR}}(I) \geq (n+1) - f(\mathcal{G}(I))$.

Proof. To obtain the target genome, the number of free cycles in $\mathcal{G}(I)$ must reach $n+1$ and, by Lemma 7, a reversal increases at most by one the number of free cycles.   □

3.2   Approximation Algorithm

In this section, we show an algorithm for the FSBR problem with approximation factor of 2 based on a reduction to the SBR problem.

Let I_{FSBR} and I_{SBR} represent instances of the FSBR and SBR problems, respectively. The idea of the algorithm consists in reducing an instance $I_{\mathrm{FSBR}} = (\pi, \breve{\pi}, \breve{\iota}^{\min}, \breve{\iota}^{\max})$ to an instance $I_{\mathrm{SBR}} = (\pi', \breve{\pi}, \breve{\iota}')$, by using a polynomial time function g, such that $\pi' = \pi$, $\breve{\pi}' = \breve{\pi}$, and the size of each intergenic region $\breve{\iota}_i'$ of the instance I_{SBR} attends the minimum $\breve{\iota}_i^{\min}$ and maximum $\breve{\iota}_i^{\max}$ sizes of its respective intergenic regions in the instance I_{FSBR}. Note that a solution for the instance $I_{\mathrm{SBR}} = g(I_{\mathrm{FSBR}})$ is also a solution for the instance I_{FSBR}, since all the constraints will be satisfied.

In the following, we describe the function g. Given an instance $I_{\mathrm{FSBR}} = (\pi, \breve{\pi}, \breve{\iota}^{\min}, \breve{\iota}^{\max})$, the function g creates an instance $I_{\mathrm{SBR}} = (\pi', \breve{\pi}, \breve{\iota}')$ as follows: (a) $\pi' = \pi$, (b) $\breve{\pi}' = \breve{\pi}$, and (c) $\breve{\iota}'$ depends of the case of the I_{FSBR} instance. Note that, by associating a weight to each gray edge in the instance I_{SBR}, we are assigning the values of $\breve{\iota}'$. Initially, for each free cycle $C = (c_1, \ldots, c_k)$ in $\mathcal{G}(I_{\mathrm{FSBR}})$, we need to find a list of non-negative integer numbers $S = (s_1, \ldots, s_k)$, such that $\sum_{i=1}^{k} s_i = W_b(C)$ and $w_g^{\min}(e_i) \leq s_i \leq w_g^{\max}(e_i)$ where e_i is the gray edge between black edges of labels c_i and $c_{i \bmod k+1}$ for $1 \leq i \leq k$, and assign to its respective gray edges in $\mathcal{G}(I_{\mathrm{SBR}})$ the corresponding values. Note that, by definition, it is always possible to find such list S considering free cycles. Besides, note that each free cycle of $\mathcal{G}(I_{\mathrm{FSBR}})$ is mapped into a balanced cycle in $\mathcal{G}(I_{\mathrm{SBR}})$, since the total weight of black and gray edges in the cycle are the same. The weights in the gray edges of the cycles in $\mathcal{G}(I_{\mathrm{SBR}})$ that correspond to unstable or auxiliary cycles in $\mathcal{G}(I_{\mathrm{FSBR}})$ depends on the case of the instance I_{FSBR}.

If it is case (i), in which the auxiliary cycles must transfer nucleotides to the unstable cycles, then for each gray edge e of an unstable cycle C in $\mathcal{G}(I_{\mathrm{FSBR}})$, we assign to its corresponding edge in $\mathcal{G}(I_{\text{SBR}})$ the value $w_g^{\min}(e)$. The same process is performed with the auxiliary cycles assuming that they are sorted in a decreasing order of $gap_{\min}$. Note that, if $\sum_{C \in \mathcal{A}} gap_{\min}(C) + \sum_{C \in \mathcal{U}} gap_{\min}(C) > 0$, the weight of some gray edges in $\mathcal{G}(I_{\text{SBR}})$ corresponding to the last auxiliary cycle C' in $\mathcal{G}(I_{\text{FSBR}})$ need to be adjusted to obtain an instance SBR, such that $\sum_{\breve{\pi}_i' \in \breve{\pi}'} \breve{\pi}_i' = \sum_{\breve{\iota}_i' \in \breve{\iota}'} \breve{\iota}_i'$. Note that, by the construction of the auxiliary set $\mathcal{A}$, $gap_{\min}(C') > \sum_{C \in \mathcal{A}} gap_{\min}(C) + \sum_{C \in \mathcal{U}} gap_{\min}(C)$, then this adjustment is always possible to be performed while not violating the minimum and maximum weight allowed in each gray edge of C'. If it is case (ii), in which the auxiliary cycles must receive nucleotides from the unstable cycles, we perform a similar process but considering the assignment of the value $w_g^{\max}(e)$ to the corresponding gray edges in the cycles of $\mathcal{G}(I_{\text{SBR}})$.

If neither case (i) nor (ii) occurs, it means that the set of auxiliary cycles $\mathcal{A}$ is empty. In this case, it is only necessary to assign the weight on the gray edges in $\mathcal{G}(I_{\text{SBR}})$ corresponding to the gray edges of the unstable cycles in $\mathcal{G}(I_{\text{FSBR}})$ respecting the minimum and maximum allowed weight in each of them. One way to do this is, for each gray edge e of an unstable cycle in $\mathcal{G}(I_{\text{FSBR}})$, to assign the weight $w_g^{\min}(e)$ to the gray edge e in $\mathcal{G}(I_{\text{SBR}})$. After that, we redistribute the weight $\sum_{C \in \mathcal{U}} W_b(C) - \sum_{C \in \mathcal{U}} W_g^{\min}(C)$ in the gray edges that will not exceed the maximum allowed weight. Performing this procedure is always possible since $\sum_{C \in \mathcal{U}} W_g^{\min}(C) \le \sum_{C \in \mathcal{U}} W_b(C) \le \sum_{C \in \mathcal{U}} W_g^{\max}(C)$.

Figure 4 shows an example of the reduction process using the g function in the instance $I_{\text{FSBR}} = (\pi = (+3\ +2\ +1\ +4\ +5), \breve{\pi} = (1,3,0,2,6,2), \breve{\iota}^{\min} = (3,2,4,0,2,1), \breve{\iota}^{\max} = (4,3,5,4,6,2))$, resulting in the instance $I_{\text{SBR}} = (\pi' = (+3\ +2\ +1\ +4\ +5), \breve{\pi}' = (1,3,0,2,6,2), \breve{\iota}' = (3,2,4,1,2,2))$. Note that we have the same cycles $C_1 = (3,1)$, $C_2 = (4,2)$, $C_3 = (5)$, and $C_4 = (6)$ in both instances. The I_{FSBR} instance represents the case (i) and we have $\mathcal{U} = \{C_1\}$, $\mathcal{S} = \{C_2, C_3, C_4\}$, $\mathcal{A} = \{C_2, C_3\}$ (considering the cycles sorted in a decreasing order by $gap_{\min}$, we have (C_3, C_2) with $gap_{\min}(C_3) = 4$ and $gap_{\min}(C_2) = 3$), and $\mathcal{F} = \{C_4\}$. The instance I_{SBR} has only one balanced cycle, which is the C_4 cycle. Note that in this example $\sum_{C \in \mathcal{A}} gap_{\min}(C) + \sum_{C \in \mathcal{U}} gap_{\min}(C) = 1$. For this reason, in the I_{SBR} instance, the gray edge $(+3, -4)$ of the cycle C_2 (which is the last auxiliary cycle) was adjusted with value 1 instead of $w_g^{\min}((+3, -4)) = 0$.

Based on the g function previously described, we obtain the following lemmas.

Lemma 9. *Given a solution S for an instance I_{SBR}, such that $I_{\text{SBR}} = g(I_{\text{FSBR}})$, S is also a solution for the instance I_{FSBR}.*

Proof. Directly by the construction of the function g. □

Lemma 10. *Using the function g to reduce an instance I_{FSBR} to an instance I_{SBR}, we have that $f(\mathcal{G}(I_{\text{FSBR}})) = b(\mathcal{G}(I_{\text{SBR}}))$.*

Proof. Observe that the cycles are the same in both instances since $\pi = \pi'$ and $\breve{\pi} = \breve{\pi}'$, except by the weight in the gray edges. Note that the free cycles in I_{FSBR} are mapped into a balanced cycle in I_{SBR}. Besides, the unstable and auxiliary cycles from I_{FSBR} are mapped into unbalanced cycles in I_{SBR}. Thus, $f(\mathcal{G}(I_{\text{FSBR}})) = b(\mathcal{G}(I_{\text{SBR}}))$ and the lemma follows. □

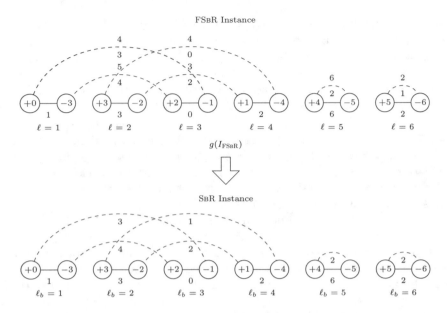

Fig. 4. Example of the reduction process of an instance I_{FSBR} (on the top) to an instance I_{SBR} (on the bottom) using the g function.

Theorem 11. *The* FSBR *problem is approximable by a factor of 2.*

Proof. Note that, by Lemmas 9 and 10, it is possible to reduce an instance I_{FSBR} to an instance I_{SBR} using a polynomial function g, such that $f(\mathcal{G}(I_{\text{FSBR}})) = b(\mathcal{G}(I_{\text{SBR}}))$ and a solution for I_{SBR} is also a solution for I_{FSBR}. This implies that using the function g, the lower bounds shown in Proposition 6 and Lemma 8 have the same value for the instances $I_{\text{SBR}} = g(I_{\text{FSBR}})$ and I_{FSBR}, respectively. I_{SBR} can be turned into a target genome using at most $2((n+1) - b(\mathcal{G}(I_{\text{SBR}})))$ reversals [10] and the theorem follows. $\qquad\square$

4   Conclusion

In this work, we investigated a generalized variant of the SORTING BY REVERSALS WITH INTERGENIC REGIONS problem, which we call the SORTING BY REVERSALS WITH FLEXIBLE INTERGENIC REGIONS problem. This variant has less strict constraints regarding the sizes of the intergenic regions. For this reason, it is possible to specify a range of acceptable values for each one of them. We showed a lower bound and presented an algorithm that guarantees an approximation factor of 2 for the problem.

As future works, it is possible to extend the model by incorporating more rearrangement events (e.g., transpositions, insertions and deletions) and considering that the genomes may not share the same set of genes.

Acknowledgments. This work was supported by the National Council of Technological and Scientific Development, CNPq (grants 140272/2020-8 and 425340/2016-3), the Coordenação de Aperfeiçoamento de Pessoal de Nível Superior - Brasil (CAPES) - Finance Code 001, and the São Paulo Research Foundation, FAPESP (grants 2013/08293-7, 2015/11937-9, 2017/12646-3, and 2019/27331-3).

References

1. Bader, D.A., Moret, B.M.E., Yan, M.: A linear-time algorithm for computing inversion distance between signed permutations with an experimental study. J. Comput. Biol. **8**, 483–491 (2001)
2. Bafna, V., Pevzner, P.A.: Genome rearrangements and sorting by reversals. SIAM J. Comput. **25**(2), 272–289 (1996)
3. Bafna, V., Pevzner, P.A.: Sorting by transpositions. SIAM J. Discrete Math. **11**(2), 224–240 (1998)
4. Biller, P., Guéguen, L., Knibbe, C., Tannier, E.: Breaking good: accounting for fragility of genomic regions in rearrangement distance estimation. Genome Biol. Evol. **8**(5), 1427–1439 (2016)
5. Biller, P., Knibbe, C., Beslon, G., Tannier, E.: Comparative genomics on artificial life. In: Beckmann, A., Bienvenu, L., Jonoska, N. (eds.) CiE 2016. LNCS, vol. 9709, pp. 35–44. Springer, Cham (2016). https://doi.org/10.1007/978-3-319-40189-8_4
6. Brito, K.L., Jean, G., Fertin, G., Oliveira, A.R., Dias, U., Dias, Z.: Sorting by genome rearrangements on both gene order and intergenic sizes. J. Comput. Biol. **27**(2), 156 174 (2020)
7. Fertin, G., Jean, G., Tannier, E.: Algorithms for computing the double cut and join distance on both gene order and intergenic sizes. Algorithms Mol. Biol. **12**(1), 16 (2017)
8. Hannenhalli, S., Pevzner, P.A.: Transforming men into mice (polynomial algorithm for genomic distance problem). In: Proceedings of the 36th Annual Symposium on Foundations of Computer Science (FOCS 1995), pp. 581–592. IEEE Computer Society Press, Washington, DC (1995)
9. Hannenhalli, S., Pevzner, P.A.: Transforming cabbage into turnip: polynomial algorithm for sorting signed permutations by reversals. J. ACM **46**(1), 1–27 (1999)
10. Oliveira, A.R., et al.: Sorting signed permutations by intergenic reversals. IEEE/ACM Trans. Comput. Biol. Bioinform. (2020)
11. Oliveira, A.R., Jean, G., Fertin, G., Brito, K.L., Dias, U., Dias, Z.: A 3.5-approximation algorithm for sorting by intergenic transpositions. In: Martín-Vide, C., Vega-Rodríguez, M.A., Wheeler, T. (eds.) AlCoB 2020. LNCS, vol. 12099, pp. 16–28. Springer, Cham (2020). https://doi.org/10.1007/978-3-030-42266-0_2

Improved DNA-versus-Protein Homology Search for Protein Fossils

Yin Yao[1,2] and Martin C. Frith[1,2,3(✉)]

[1] Graduate School of Frontier Sciences, University of Tokyo, Chiba, Japan
mcfrith@edu.k.u-tokyo.ac.jp
[2] Artificial Intelligence Research Center, AIST, Tokyo, Japan
[3] Computational Bio Big-Data Open Innovation Laboratory (CBBD-OIL), AIST, Tokyo, Japan

Abstract. Protein fossils, i.e. noncoding DNA descended from coding DNA, arise frequently from transposable elements (TEs), decayed genes, and viral integrations. They can reveal, and mislead about, evolutionary history and relationships. They have been detected by comparing DNA to protein sequences, but current methods are not optimized for this task. We describe a powerful DNA-protein homology search method. We use a 64×21 substitution matrix, which is fitted to sequence data, automatically learning the genetic code. We detect subtly homologous regions by considering alternative possible alignments between them, and calculate significance (probability of occurring by chance between random sequences). Our method detects TE protein fossils much more sensitively than `blastx`, and $>10\times$ faster. Of the $\sim$7 major categories of eukaryotic TE, three were long thought absent in mammals: we find two of them in the human genome, polinton and DIRS/Ngaro. This method increases our power to find ancient fossils, and perhaps to detect non-standard genetic codes. The alternative-alignments and significance paradigm is not specific to DNA-protein comparison, and could benefit homology search generally.

Keywords: Sequence alignment · Frameshift · Pseudogene · Homology

1 Introduction

Genomes are littered with protein fossils, old and young. They can be found by comparing DNA to known proteins: new transposable element (TE) families have been discovered in this way [26]. An interesting class of protein fossils comes from ancient integrations of viral DNA into genomes, enabling the field of paleovirology [17]. The DNA sequences of protein fossils often have similarity to distantly-related genomes (e.g. mammal versus fish), simply because the parent gene evolved slowly, so it is important to know that they are protein fossils in order to understand this similarity [29]. DNA-protein homology search is also used to classify DNA reads from unknown microbes, including nanopore and PacBio reads with many sequencing errors [16]. DNA-protein comparison can

© Springer Nature Switzerland AG 2021
C. Martín-Vide et al. (Eds.): AlCoB 2021, LNBI 12715, pp. 146–158, 2021.
https://doi.org/10.1007/978-3-030-74432-8_11

be used to find frameshifts during evolution of functional proteins [27], and programmed ribosomal frameshifts [37]. A more specialized and complex kind of DNA-protein comparison, outside this study's scope, considers introns and other gene features to identify genes.

DNA-protein homology search is a classical problem with many old solutions [4,11–13,15,20,22–24,32,36,41]. A notable one is "three-frame alignment" [41], which we believe is the simplest and fastest reasonable way to do frameshifting DNA-protein alignment. Nevertheless, we can significantly improve DNA-protein homology search in these aspects:

- Better parameters for the (dis)favorability of substitutions, deletions, insertions, and frameshifts. Most previous methods use standard parameters such as the BLOSUM62 substitution matrix, which is designed for functional proteins, and likely completely inappropriate for protein fossils. We optimize these parameters by fitting them to sequence data.
- Instead of a 20×20 substitution matrix, use a 64×21 matrix (64 codons $\times$ 20 amino acids plus STOP). This allows e.g. preferred alignment of asparagine (which is encoded by aac and aat) to agc than to tca, which both encode serine.
- Incorporate frameshifts into affine gaps. Because gaps are somewhat rare but often long, it is standard to disfavor opening a gap more than extending a gap. However, most previous methods favor frameshifts equally whether isolated or contiguous with a longer gap.
- Detect homologous regions based on not just one alignment between them, but on many possible alternative alignments. This is expected to detect subtle homology more powerfully [1,6].
- Calculate significance, i.e. the probability of such a strong similarity occurring by chance between random sequences. To this day, for ordinary alignment, BLAST can only calculate significance for a few hardcoded sets of substitution and gap parameters. We can do it for any parameters, for similarities based on many alternative alignments.

We also aimed for maximum simplicity and speed, inspired by three-frame alignment.

2   Methods

2.1   Alignment Elements

We define a DNA-protein alignment to consist of: matches (3 bases aligned to 1 amino acid), base insertions, and base deletions. To keep things simple, insertions are not allowed between bases aligned to one amino acid. A deletion of length not divisible by 3 leaves "dangling" bases (Fig. 1): for simplicity, we do not attempt to align these (equivalently, align them to the amino acid with score 0).

```
Ser-TyrAlaThrMetLeuTrpAspGln--Leu***
tctCtat---acg--cctctga-atcagCAttctaa
```

Fig. 1. Example of a DNA-versus-protein alignment. *** indicates a protein end from translation of a stop codon. Insertions are bold uppercase. "Dangling" bases, left by deletions of length not divisible by 3, are underlined gray.

2.2 Scoring Scheme

An alignment's score is the sum of:

- Score for aligning amino acid u to base triplet V: S_{uV}

- Score for an insertion of k bases: $a_I + b_I k + \begin{cases} 0 & \text{if } k \bmod 3 = 0 \\ f_I & \text{if } k \bmod 3 = 1 \\ g_I & \text{if } k \bmod 3 = 2 \end{cases}$

- Score for a deletion of k bases: $a_D + b_D k + \begin{cases} 0 & \text{if } k \bmod 3 = 0 \\ f_D & \text{if } k \bmod 3 = 1 \\ g_D & \text{if } k \bmod 3 = 2 \end{cases}$

This scheme extravagantly uses 4 frameshift parameters (f_I, g_I, f_D, g_D), because it's based on a probability model with 4 frameshift transitions (Fig. 2), and we can't think of a good way to simplify the model. Overall, our alignment scheme is similar to FramePlus [13] and especially to aln [11].

2.3 Finding a Maximum-Score Local Alignment

A basic approach is to find an alignment with maximum possible score, between any parts of a protein sequence $R_1 \ldots R_M$ and a DNA sequence $q_1 \ldots q_N$. Let Q_j mean the triplet q_{j-2}, q_{j-1}, q_j. We can calculate the maximum possible score X_{ij} for any alignment ending just after R_i and q_j, for $0 \le i \le M$ and $0 \le j \le N$ (with notation y/Y for deletion and z/Z for insertion):

$$y_1 = Y_{i-1\ j-2} + [b_D + f_D] \qquad\qquad z_1 = Z_{i\ j-1} + [b_I + f_I]$$
$$y_2 = Y_{i-1\ j-1} + [2b_D + g_D] \qquad\quad z_2 = Z_{i\ j-2} + [2b_I + g_I]$$
$$y_3 = Y_{i-1\ j} + [3b_D] \qquad\qquad\qquad z_3 = Z_{i\ j-3} + [3b_I]$$
$$X_{ij} = \max(X_{i-1\ j-3} + S_{R_iQ_j},\ y_1,\ y_2,\ y_3,\ z_1,\ z_2,\ z_3,\ 0)$$
$$Y_{ij} = \max(X_{ij} + a_D,\ y_3) \qquad\qquad Z_{ij} = \max(X_{ij} + a_I,\ z_3)$$

The boundary condition is: if $i < 0$ or $j < 0$, $X_{ij} = Y_{ij} = Z_{ij} = -\infty$ (which takes care of $R_{i<1}$ and $Q_{j<3}$). The maximum possible alignment score is $\max(X_{ij})$, and an alignment with this score can be found by a standard traceback [5].

For each (i, j) this algorithm retrieves 7 previous results, and performs 9 pairwise maximizations and 9 additions (which could be reduced to 6 additions if each insertion cost equals its corresponding deletion cost). This is slightly slower than three-frame alignment, which retrieves 5 previous results and performs 7 pairwise maximizations and 6 additions.

2.4 Probability Model

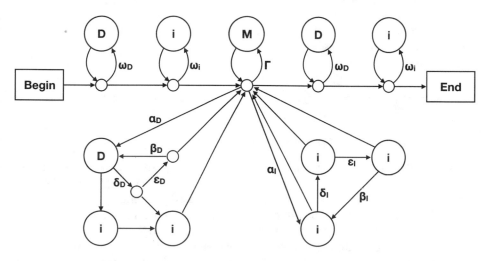

Fig. 2. A probability model for related DNA and protein sequences. The arrows are labeled with probabilities of traversing them. Each pass through an **i** state generates one base $v \in \{\mathsf{a}, \mathsf{c}, \mathsf{g}, \mathsf{t}\}$, with probabilities ψ_v. Each pass through a **D** state generates one amino acid u, with probabilities ϕ_u. Each pass through the **M** state generates one amino acid u aligned to three bases $V = v_1 v_2 v_3$, with probabilities π_{uV}. The two bottom-left **i** states correspond to "dangling" bases.

The preceding algorithm is equivalent to finding a maximum-probability path generating the sequences, through a probability model (Fig. 2). The score and model parameters are related like this:

$$S'_{uV} = \exp\left(\frac{S_{uV}}{t}\right) = \frac{\Gamma}{\omega_D \omega_i^3} \cdot \frac{\pi_{uV}}{\phi_u \psi_V}$$

$$a'_I = \exp\left(\frac{a_I}{t}\right) = \frac{\alpha_I(1 - \beta_I)}{\beta_I} \qquad a'_D = \exp\left(\frac{a_D}{t}\right) = \frac{\alpha_D(1 - \beta_D)}{\beta_D}$$

$$b'_I = \exp\left(\frac{b_I}{t}\right) = \frac{\sqrt[3]{\beta_I \delta_I \epsilon_I}}{\omega_i} \qquad b'_D = \exp\left(\frac{b_D}{t}\right) = \sqrt[3]{\frac{\beta_D \delta_D \epsilon_D}{\omega_D}}$$

$$f'_I = \exp\left(\frac{f_I}{t}\right) = \frac{1 - \delta_I}{1 - \beta_I} \sqrt[3]{\frac{\beta_I^2}{\delta_I \epsilon_I}} \qquad f'_D = \exp\left(\frac{f_D}{t}\right) = \frac{1 - \delta_D}{1 - \beta_D} \sqrt[3]{\frac{\beta_D^2}{\delta_D \epsilon_D \omega_D^2}} \Big/ \omega_i^2$$

$$g'_I = \exp\left(\frac{g_I}{t}\right) = \frac{1 - \epsilon_I}{1 - \beta_I} \sqrt[3]{\frac{\beta_I \delta_I}{\epsilon_I^2}} \qquad g'_D = \exp\left(\frac{g_D}{t}\right) = \frac{1 - \epsilon_D}{1 - \beta_D} \sqrt[3]{\frac{\beta_D \delta_D}{\epsilon_D^2 \omega_D}} \Big/ \omega_i$$

Here ψ_V is defined to be $\psi_{v_1} \psi_{v_2} \psi_{v_3}$, and t is an arbitrary positive constant (because multiplying all the score parameters by a constant makes no difference to alignment). An alignment score is then: $t \ln[\text{prob}(\text{path} \& \text{sequences})/\text{prob}(\text{null path} \& \text{sequences})]$, where a "null path" is a path that never traverses the Γ, α_D, or α_I arrows [10].

Balanced Length Probability. A fundamental property of local alignment models is whether they are biased towards longer or shorter alignments [10]. If ω_D and ω_i are large (close to 1) and $\Gamma + \alpha_D + \alpha_I$ is small, there is a bias in favor of shorter alignments. In the converse situation, there is a bias towards longer alignments. It can be shown (using the method of [10]) that our DNA-protein model is unbiased when

$$\frac{\Gamma}{\omega_D \omega_i^3} + \frac{a_I' b_I'(f_I' + g_I' b_I' + b_I'^2)}{1 - b_I'^3} + \frac{a_D' b_D'(f_D' + g_D' b_D' + b_D'^2)}{1 - b_D'^3} = 1. \tag{1}$$

2.5   Sum over All Alignments Passing Through (i, j)

To find subtly homologous regions, we should assess their homology without fixing an alignment [1,6]. In other words, we should use a homology score like this: $t \ln \left[\sum_{\text{paths}} \text{prob(path \& sequences)} / \text{prob(null path \& sequences)} \right]$. However, if the sum is taken over all possible paths, we learn nothing about location of the homologous regions, which is important if e.g. the DNA sequence is a chromosome. There is a kind of uncertainty principle here: the more we pin down the alignment, the less power we have to detect homology. As a compromise, we sum over all paths passing through one (protein, DNA) coordinate pair (i, j). This has two further benefits: it is approximated by the seed-and-extend search used for big sequence data, and we can calculate significance.

To calculate this sum over paths, we first run a Forward algorithm for $0 \leq i \leq M$ and $0 \leq j \leq N$:

$$y_1 = [b_D' f_D'] Y_{i-1\ j-2}^F \qquad y_2 = [b_D'^2 g_D'] Y_{i-1\ j-1}^F \qquad y_3 = [b_D'^3] Y_{i-1\ j}^F$$
$$z_1 = [b_I' f_I'] Z_{i\ j-1}^F \qquad z_2 = [b_I'^2 g_I'] Z_{i\ j-2}^F \qquad z_3 = [b_I'^3] Z_{i\ j-3}^F$$
$$X_{ij}^F = S_{R_i Q_j}' X_{i-1\ j-3}^F + y_1 + y_2 + y_3 + z_1 + z_2 + z_3 + 1$$
$$Y_{ij}^F = a_D' X_{ij}^F + y_3 \qquad Z_{ij}^F = a_I' X_{ij}^F + z_3$$

The boundary condition is: if $i < 0$ or $j < 0$, $X_{ij}^F = Y_{ij}^F = Z_{ij}^F = 0$. We then run a Backward algorithm for $M \geq i \geq 0$ and $N \geq j \geq 0$:

$$y_1 = [b_D' f_D'] Y_{i+1\ j+2}^B \qquad y_2 = [b_D'^2 g_D'] Y_{i+1\ j+1}^B \qquad y_3 = [b_D'^3] Y_{i+1\ j}^B$$
$$z_1 = [b_I' f_I'] Z_{i\ j+1}^B \qquad z_2 = [b_I'^2 g_I'] Z_{i\ j+2}^B \qquad z_3 = [b_I'^3] Z_{i\ j+3}^B$$
$$X_{ij}^B = S_{R_{i+1} Q_{j+3}}' X_{i+1\ j+3}^B + y_1 + y_2 + y_3 + z_1 + z_2 + z_3 + 1$$
$$Y_{ij}^B = a_D' X_{ij}^B + y_3 \qquad Z_{ij}^B = a_I' X_{ij}^B + z_3$$

The boundary condition is: if $i > M$ or $j > N$, $X_{ij}^B = Y_{ij}^B = Z_{ij}^B = 0$. Finally, $t \ln[X_{ij}^F X_{ij}^B]$ is the desired homology score, for all paths passing through (i, j).

2.6   Significance Calculation

The just-described homology score is similar to that of "hybrid alignment", which has a conjecture regarding significance [39]. (Hybrid alignment sums over paths ending at (i, j), instead of passing through (i, j).) We make a similar conjecture. Suppose we compare a random i.i.d. protein sequence of length M and letter probabilities Φ_u to a random i.i.d. DNA sequence of length N and triplet probabilities Ψ_V. We conjecture that the score $s_{\max} = t \ln[\max_{ij}(X_{ij}^F X_{ij}^B)]$ follows a Gumbel distribution:

$$\text{prob}(s_{\max} < s) = \exp(-KMNe^{-s/t}), \tag{2}$$

in the limit that M and N are large, provided that:

$$\left(\sum_{u,V} \Phi_u \Psi_V S'_{uV}\right) + \frac{a'_I b'_I (f'_I + g'_I b'_I + b'^2_I)}{1 - b'^3_I} + \frac{a'_D b'_D (f'_D + g'_D b'_D + b'^2_D)}{1 - b'^3_D} = 1. \tag{3}$$

Eq. 3 is analogous to Eq. 27 or 28 in [39], see also [10]. In practice, we assume that $\Phi_u = \phi_u$ and $\Psi_V = \psi_V$, which makes Eq. 3 equivalent to Eq. 1.

This conjecture leaves one unknown Gumbel parameter K. We estimate it by brute-force simulation of 50 pseudorandom sequence pairs [40], with $\Phi_u = \phi_u$, $\Psi_V = \psi_V$, $M = 200$ and $N = 602$, which takes zero human-perceptible run time.

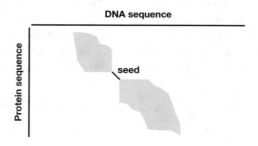

Fig. 3. Sketch of seed-and-extend heuristic for homology search.

2.7 Seed-and-Extend Heuristic

To find homologous regions in big sequence data, we use a BLAST-like seed-and-extend heuristic (Fig. 3) [2]. We first find "seeds": we currently use exact-matches (via the genetic code), which can be sensitive if short, but we could likely get better sensitivity per run time with inexact seeds [28,33]. Our seeds have variable length: starting from each DNA base, we get the shortest seed that occurs $\leq m$ times in the protein data [19]. These seeds have no score threshold. We then try a gapless X-drop extension in both directions, and if the score achieves a threshold d, we try a "Forward" extension in both directions.

We use our Forward algorithm, modified for semi-global instead of local alignment. In each direction, we sum over alignments starting at the seed and ending anywhere: thus the algorithm's +1 is done only at the first (i, j) next to the seed, and we accumulate the sum $W = \sum_{ij} X_{ij}^F$. We run this algorithm in increasing order of antidiagonal $(3i + j)$ on the seed's right side (decreasing order on the left side). If X_{ij}^F is less than a fraction f of W accumulated over previous antidiagonals, we stop extending, which defines the boundary of the gray region in Fig. 3. The final homology score is $t \ln[W_{\text{left}}] + \text{seed score} + t \ln[W_{\text{right}}]$.

Sum-of-path algorithms are prone to numerical overflow [5]. To prevent that: once per 32 antidiagonals, we multiply all the X^F, Y^F, and Z^F values in the last six antidiagonals by a scaling factor of $1/W$.

A score with no alignment is disconcerting, so we get a representative alignment by a similar semi-global modification of our maximum-score alignment algorithm. To avoid redundancy, we prioritize homology predictions by score (breaking ties arbitrarily), and discard any prediction whose representative alignment shares an (i, j) left or right end with a higher-priority prediction.

2.8   Fitting Substitution and Gap Parameters to Sequence Data

We can fit the parameters to some related (unaligned) DNA and proteins, by an iterative Baum-Welch algorithm [5]. We implemented two versions of this: an exact $O(MN)$ version, and a seed-and-extend version. The seed-and-extend version, at each iteration, finds significantly homologous regions (with -K1 filtering, see below) and gets expected counts from the seeds and extend regions (gray areas in Fig. 3). It does not infer ϕ_u, ψ_v, ω_D, or ω_i in the usual way: at each iteration, it sets $\phi_u = \sum_V \pi_{uV}$, $\psi_v = \sum_{uij}(\pi_{u\,vij} + \pi_{u\,ivj} + \pi_{u\,ijv})/3$, and $\omega_D = \omega_i^3 =$ the value that satisfies Eq. 1 (found by bisection with bounds $1 > \omega_i^3 > \beta_I \delta_I \epsilon_I$ and $1 > \omega_D > \beta_D \delta_D \epsilon_D$). We set $t = 3/\ln[2]$ to get scores in third-bit units.

3   Results

3.1   Software

The $O(MN)$ alignment and fitting code is available at https://github.com/Yao-Yin/protein-dna-align-EM. Seed-and-extend fitting and homology search, and estimation of K by full Forward-Backward algorithm, are available in LAST (https://gitlab.com/mcfrith/last).

3.2   Parameter Fitting

We applied our $O(MN)$ fitting to a set of human processed pseudogenes and their parent proteins from Pseudofam [21]. To avoid bias, we began the first iteration with $\pi_{uV} = 1/(21 \cdot 64)$. The fitting discovered the genetic code: for each codon V, its encoded amino acid has maximum S_{uV}.

Sometimes, our fitting had an undesirable feature: the S_{uV} values for some cg-containing codons were all negative. This is presumably due to the well-known depletion of cg in human DNA, which can be captured in π_{uV} but not ψ_v. As an ad hoc fix, we set $\psi_V = \sum_u \pi_{uV}$ (after $O(MN)$ fitting, and at each iteration of seed-and-extend fitting).

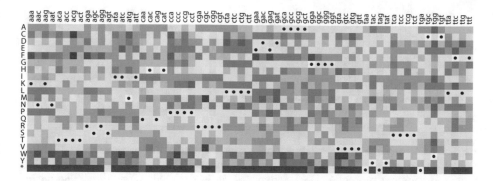

Fig. 4. Substitution matrix inferred from human chromosome 21 versus RepeatMasker proteins. Darker red means more disfavored and paler yellow means more favored. Black dots indicate the standard genetic code. (Color figure online)

Next, we applied our seed-and-extend fitting to human chromosome 21 (hg38 chr21) and transposable element (TE) proteins from RepeatMasker 4.1.0 [30]. The result primarily favors genetic-code matches (Fig. 4), and secondarily favours single a↔g or c↔t mismatches, e.g. asparagine scores +5 with agc and −14 with tca. The gap scores are $a_I, b_I, f_I, g_I = -28, -1, +3, 0$ and $a_D, b_D, f_D, g_D = -23, -1, +3, 0$. So frameshifts are not disfavored, perhaps because RepeatMasker's proteins are close to the fossils's most recent active ancestors. The positive $f_{I,D}$ values might be caused by the gap-length distribution not fitting the simple affine model, with an excess of length-1 and length-4 gaps [35].

3.3 Significance Calculation and Simple Sequences

To test the accuracy of our significance estimates, for the chr21-TE parameters, we calculated s_{max} by our full Forward-Backward algorithm for 1000 pairs of random i.i.d. protein and codon sequences, with $\Phi_u = \phi_u$, $\Psi_V = \psi_V$, $M = 200$, and $N = 602$. The observed distribution of s_{max} agrees reasonably well with that predicted by Eq. 2 (Fig. 5A).

To test whether our significance estimates apply to our seed-and-extend homology search, we compared one pair of random i.i.d. protein and DNA sequences, with $\Phi_u = \phi_u$, $\Psi_v = \psi_v$, and lengths equal to the number of unambiguous letters in the TE proteins and chr21. The search sensitivity depends on the seed parameter m: as m increases, sensitivity increases, and the distribution of homology scores approaches the Gumbel prediction (Fig. 5B).

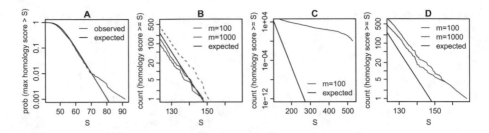

Fig. 5. Distributions of homology scores using the chr21-TE parameters. (**A**) Exact homology scores for random protein & codon sequences. (**B**) Seed-and-extend homology scores for random protein & DNA sequences. m is the seed count limit. The dashed line shows a test with $(\Phi, \Psi) \neq (\phi, \psi)$. (**C**) Seed-and-extend homology scores for TE proteins & reversed chr21 without and (**D**) with simple-sequence masking.

We then considered $(\Phi, \Psi) \neq (\phi, \psi)$, because the marginal frequencies of π_{uV} differ from the letter abundances in the TE proteins and chr21, e.g. $\psi_a{:}\psi_c{:}\psi_g{:}\psi_t$ = 40:19:18:23 but chr21 is 29:21:21:29. So we compared another pair of random i.i.d. protein and DNA sequences, with Φ_u and Ψ_v equal to the frequencies in the TE proteins and chr21. In this test, the E-values (expected counts) were too low by a factor of about 3 (Fig. 5B).

Homology search is confounded by "simple sequences", e.g. `ttttcttt` `tttcctt`, which evolve frequently and independently. There are various methods to suppress such false homologies, but most do not fully succeed [8,9]. To illustrate, we compared reversed (but not complemented) chr21 to the TE proteins: this test has no true homologies, but we found many highly-significant homology scores (Fig. 5C). Our solution is to mask the DNA and protein with `tantan` [9], which eliminates extremely-significant false homologies, at least in this test (Fig. 5D). Further testing is warranted, e.g. here we used a default `tantan` parameter $r = 0.005$, but 0.02 was suggested for DNA-protein comparison [9].

3.4   Comparison to Blastx

To test whether our homology search is more sensitive than standard methods, we compared chr21 to the TE proteins with NCBI BLAST 2.11.0:

```
makeblastdb -in RepeatPeps.lib -dbtype prot -out DB
blastx -query chr21.fa -db DB -evalue 0.1 -outfmt 7 > out
```

We repeated this comparison with our method (in LAST version 1177):

```
lastdb -q -c -R01 myDB RepeatPeps.lib
last-train --codon -X1 myDB chr21.fa > train.out
lastal -p train.out -D1e9 -m100 -K1 myDB chr21.fa > out
```

Option `-q` appends `*` to each protein; `-R01` lowercases simple sequence with `tantan`; `-c` requests masking of lowercase; `-X1` treats matches to unknown residues (which are frequent in these proteins) as neutral instead of disfavored; -D1e9 sets the significance threshold to 1 random hit per 10^9 basepairs; -m100 sets $m = 100$; -K1 omits alignments whose DNA range lies in that of a higher-scoring alignment.

This test indicated that our method has much better sensitivity and speed. The single-threaded runtimes were 193 min for blastx and 18 min for lastal. (We also tried fasty from FASTA version 36.3.8h [23], but it didn't finish within 30 h.) blastx found alignments at 2604 non-overlapping sites on the two strands of chr21, of which all but 23 overlapped LAST alignments. LAST found alignments at 6640 non-overlapping sites, of which 4499 did not overlap blastx alignments. All but 21 of LAST's sites overlapped same-strand annotations by RepeatMasker open-4.0.6 - Dfam 2.0 (excluding Simple_repeat and Low_complexity) [30,34], suggesting they are not spurious. Note that Repeat-Masker finds TEs by DNA models of tightly-defined TE families, which is likely superior to DNA-protein comparison when such models are available. Our approach cannot find the many TEs, such as Alu elements, that have never encoded proteins.

3.5   Discovery of Missing TE Orders in the Human Genome

Eukaryotic TEs have immense diversity, but can be classified into ~7 major orders: LTR, LINE, and tyrosine-recombinase (YR) retrotransposons, and DDE transposons, cryptons, helitrons, and polintons [38]. Three of them (YR retrotransposons, cryptons, polintons) were long thought absent in mammals [3,25].

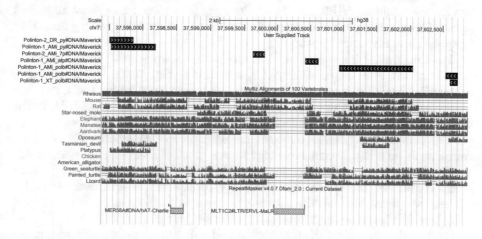

Fig. 6. An ancient polinton in human chromosome 7. Black bars: alignments to polinton proteins, arrows indicate +/− strand. Green: alignments to other vertebrate genomes [14]. Screen shot from http://genome.ucsc.edu [18]. (Color figure online)

By comparing the whole human genome (hg38) to RepeatMasker's TE proteins, with the same lastal options, we found two of these missing TE orders:

YR retrotransposons and polintons. We found polinton alignments at 18 non-overlapping genome sites, with E-values as low as 1.8e−36. These alignments covered 75–1103 bases, and seemingly-random parts of several polinton proteins (e.g. POLB, ATPase, PY, integrase). Five of the 18 sites are clustered in chromosome 7, indicating that an ancient polinton was fragmented by insertion of an LTR element (Fig. 6). Whole-genome alignments (Fig. 6, green) indicate that the LTR inserted in an ancestor of Euarchontoglires (primates and rodents), and the polinton in an ancestor of all amniotes (mammals and reptiles). A recent study also found this human locus, by searching with proteins of adintovirus, a polinton-like virus [31]. We found both major superfamilies of YR retrotransposon: DIRS alignments (covering 73–498 bases) at 20 non-overlapping sites with min E-value 3e−45, and Ngaro (128–275 bases) at 4 non-overlapping sites with min E-value 5.1e−14. The lastal CPU time scales linearly with DNA length: this search took 175 min (1619 min CPU time with chromosomes searched in parallel).

4   Discussion

Our DNA-protein homology search method seems to be fast, specific, and highly sensitive. Our reversed-sequence test found false-positive homologies with E-values $\gtrsim 0.01$ (Fig. 5D), suggesting that E-values much below this likely indicate true homologies. Note that true homology may still lead to false-positive inference, e.g. some TEs carry fragments of non-related TEs [34]. With due care, our method should enable discovery of more ancient and subtle fossils, such as the human polinton, DIRS and Ngaro elements found here. So almost all known major TE categories have left traces in the human genome, suggesting an ability to spread broadly among eukaryotes.

Possible future improvements include better seeding, and using position-specific information on variability of a sequence family [5,40]. Our significance calculation becomes inaccurate for short sequences, so a finite size correction would be useful [39]. Our method's parameter-fitting makes it versatile, but it would be better to use different parameters for fossils of different ages.

The sum-of-paths and significance paradigm is not specific to DNA-protein comparison, so could benefit homology search generally. A previous study made similar conjectures on significance of probabilistic homology scores [7]. That study considered the maximum-probability path and the sum over all paths; whereas we (following [39]) use a sum over *some* paths (emanating from an (i, j) point), with a balance condition (Eq. 3). We suspect the conjectures in [7] may be too broad: e.g. one set of substitution and gap scores corresponds to a range of probability models with different values of t [10], but only one t can appear in the Gumbel formula (Eq. 2).

Acknowledgments. We thank the Frith and Asai lab members for discussions that clarified our thinking.

References

1. Allison, L., Wallace, C.S., Yee, C.N.: Finite-state models in the alignment of macro-molecules. J. Mol. Evol. **35**(1), 77–89 (1992)
2. Altschul, S.F., et al.: Gapped BLAST and PSI-BLAST: a new generation of protein database search programs. Nucleic acids Res. **25**(17), 3389–3402 (1997)
3. Campbell, S., Aswad, A., Katzourakis, A.: Disentangling the origins of virophages and polintons. Curr. Opin. Virol. **25**, 59–65 (2017)
4. Csűrös, M., Miklós, I.: Statistical alignment of retropseudogenes and their functional paralogs. Mol. Biol. Evol. **22**(12), 2457–2471 (2005)
5. Durbin, R., Eddy, S., Krogh, A., Mitchison, G.: Biological Sequence Analysis: Probabilistic Models of Proteins and Nucleic Acids. Cambridge University Press, Cambridge (1998)
6. Eddy, S.R.: A new generation of homology search tools based on probabilistic inference. Genome Inform. **23**(1), 205–211 (2009)
7. Eddy, S.R.: A probabilistic model of local sequence alignment that simplifies statistical significance estimation. PLoS Comput. Biol. **4**(5), e1000069 (2008)
8. Frith, M.C.: Gentle masking of low-complexity sequences improves homology search. PLoS One **6**(12), e28819 (2011)
9. Frith, M.C.: A new repeat-masking method enables specific detection of homologous sequences. Nucleic Acids Res. **39**(4), e23–e23 (2011)
10. Frith, M.C.: How sequence alignment scores correspond to probability models. Bioinformatics **36**(2), 408–415 (2020)
11. Gotoh, O.: Homology-based gene structure prediction: simplified matching algorithm using a translated codon (tron) and improved accuracy by allowing for long gaps. Bioinformatics **16**(3), 190–202 (2000)
12. Guan, X., Uberbacher, E.C.: Alignments of DNA and protein sequences containing frameshift errors. Comput. Appl. Biosci. **12**(1), 31–40 (1996)
13. Halperin, E., Faigler, S., Gill-More, R.: FramePlus: aligning DNA to protein sequences. Bioinformatics **15**(11), 867–873 (1999)
14. Harris, R.S.: Improved pairwise alignment of genomic DNA. Ph.D. thesis, The Pennsylvania State University (2007)
15. Huang, X., Zhang, J.: Methods for comparing a DNA sequence with a protein sequence. Bioinformatics **12**(6), 497–506 (1996)
16. Huson, D.H., et al.: MEGAN-LR: new algorithms allow accurate binning and easy interactive exploration of metagenomic long reads and contigs. Biol. Direct **13**(1), 6 (2018)
17. Katzourakis, A., Gifford, R.J.: Endogenous viral elements in animal genomes. PLoS Genet. **6**(11), e1001191 (2010)
18. Kent, W.J., et al.: The human genome browser at UCSC. Genome Res. **12**(6), 996–1006 (2002)
19. Kiełbasa, S.M., Wan, R., Sato, K., Horton, P., Frith, M.C.: Adaptive seeds tame genomic sequence comparison. Genome Res. **21**(3), 487–493 (2011)
20. Ko, P., Narayanan, M., Kalyanaraman, A., Aluru, S.: Space-conserving optimal DNA-protein alignment. In: Proceedings of the 2004 IEEE Computational Systems Bioinformatics Conference, 2004. CSB 2004, pp. 80–88. IEEE (2004)
21. Lam, H.Y., et al.: Pseudofam: the pseudogene families database. Nucleic Acids Res. **37**(suppl_1), D738–D743 (2009)
22. Lysholm, F.: Highly improved homopolymer aware nucleotide-protein alignments with 454 data. BMC Bioinform. **13**(1), 230 (2012)

23. Pearson, W.R., Wood, T., Zhang, Z., Miller, W.: Comparison of DNA sequences with protein sequences. Genomics **46**(1), 24–36 (1997)
24. Peltola, H., Söderlund, H., Ukkonen, E.: Algorithms for the search of amino acid patterns in nucleic acid sequences. Nucleic Acids Res. **14**(1), 99–107 (1986)
25. Poulter, R.T., Butler, M.I.: Tyrosine recombinase retrotransposons and transposons. In: Mobile DNA III, pp. 1271–1291 (2015)
26. Pritham, E.J., Feschotte, C.: Massive amplification of rolling-circle transposons in the lineage of the bat Myotis lucifugus. Proc. Nat. Acad. Sci. **104**(6), 1895–1900 (2007)
27. Raes, J., Van de Peer, Y.: Functional divergence of proteins through frameshift mutations. Trends Genet. **21**(8), 428–431 (2005)
28. Roytberg, M., et al.: On subset seeds for protein alignment. IEEE/ACM Trans. Comput. Biol. Bioinform. **6**(3), 483–494 (2009)
29. Sheetlin, S.L., Park, Y., Frith, M.C., Spouge, J.L.: Frameshift alignment: statistics and post-genomic applications. Bioinformatics **30**(24), 3575–3582 (2014)
30. Smit, A., Hubley, R., Green, P.: RepeatMasker open-4.0 (2013–2015). http://www.repeatmasker.org
31. Starrett, G.J., et al.: Adintoviruses: a proposed animal-tropic family of midsize eukaryotic linear dsDNA (MELD) viruses. Virus Evol. (2020). veaa055
32. States, D., Botstein, D.: Molecular sequence accuracy and the analysis of protein coding regions. Proc. Nat. Acad. Sci. U.S.A. **88**(13), 5518 (1991)
33. Steinegger, M., Söding, J.: MMseqs2 enables sensitive protein sequence searching for the analysis of massive data sets. Nat. Biotechnol. **35**(11), 1026–1028 (2017)
34. Storer, J., Hubley, R., Rosen, J., Wheeler, T.J., Smit, A.F.: The Dfam community resource of transposable element families, sequence models, and genome annotations. Mobile DNA **12**(1), 1–14 (2021)
35. Tanay, A., Siggia, E.D.: Sequence context affects the rate of short insertions and deletions in flies and primates. Genome Biol. **9**(2), R37 (2008)
36. Tzou, P.L., Huang, X., Shafer, R.W.: NucAmino: a nucleotide to amino acid alignment optimized for virus gene sequences. BMC Bioinform. **18**(1), 138 (2017)
37. Wang, R., Xiong, J., Wang, W., Miao, W., Liang, A.: High frequency of +1 programmed ribosomal frameshifting in Euplotes octocarinatus. Sci. Rep. **6**, 21139 (2016)
38. Wells, J.N., Feschotte, C.: A field guide to eukaryotic transposable elements. Ann. Rev. Genet. **54**, 539–561 (2020)
39. Yu, Y.K., Hwa, T.: Statistical significance of probabilistic sequence alignment and related local hidden Markov models. J. Comput. Biol. **8**(3), 249–282 (2001)
40. Yu, Y.K., Bundschuh, R., Hwa, T.: Hybrid alignment: high-performance with universal statistics. Bioinformatics **18**(6), 864–872 (2002)
41. Zhang, Z., Pearson, W.R., Miller, W.: Aligning a DNA sequence with a protein sequence. J. Comput. Biol. **4**(3), 339–349 (1997)

The Maximum Weight Trace Alignment Merging Problem

Paul Zaharias[ID], Vladimir Smirnov[ID], and Tandy Warnow[✉][ID]

Department of Computer Science, University of Illinois, 201 N. Goodwin Avenue,
Urbana, IL 61801, USA
{zaharias,smirnov3,warnow}@illinois.edu

Abstract. The Maximum Weight Trace (MWT) is an optimization
problem for multiple sequence alignment that takes a set of sequences
and weights on pairs of letters from different sequences and seeks a multiple sequence alignment that maximizes the sum of the weights for the
pairs of letters that appear in the same column. MWT was introduced
by Kececioglu in 1993, then proven to be NP-hard, and heuristics and
exact solutions for MWT developed. Unfortunately none of the MWT
methods are scalable to even moderate-sized datasets. Here we propose
the MWT-AM problem (MWT for Alignment Merging), an extension of
the MWT problem to be used in a divide-and-conquer setting, where we
seek a merged alignment of a set of disjoint alignments that optimizes
the MWT score. We present variations of GCM (the Graph Clustering
Merger, originally developed for the MAGUS multiple sequence alignment method) that are specifically designed for MWT-AM. We show
that the best of these variants, which we refer to as GCM-MWT, perform well for the MWT-AM criterion. We explore GCM-MWT in comparison to other methods for merging alignments, T-coffee and MAFFT–
merge, and find that GCM-MWT produces more accurate merged alignments. GCM-MWT is available in open source form at https://github.
com/vlasmirnov/MAGUS.

Keywords: Multiple sequence alignment · Maximum Weight Trace ·
Clustering

1 Introduction

The Maximum Weight Trace (MWT) problem, introduced by Kececioglu nearly
30 years ago [5], is a classical problem in the multiple sequence alignment literature. The input to the MWT problem is a set of sequences (e.g., DNA sequences)
and weights on pairs of letters from the different sequences, and the objective is
an MSA that has the maximum total possible weight (defined to be the sum of
the weights of aligned letters in the output MSA). Kececioglu [5] showed that
MWT is NP-hard, and can be exactly solved in $O(2^B L^B B^2)$ time for B sequences

Supported by the University of Illinois.

© Springer Nature Switzerland AG 2021
C. Martín-Vide et al. (Eds.): AlCoB 2021, LNBI 12715, pp. 159–171, 2021.
https://doi.org/10.1007/978-3-030-74432-8_12

of total length L using dynamic programming. He also proposed a more practical branch-and-bound refinement of this algorithm, but the reduction in time is still not sufficient to scale to even moderate-sized datasets. Other attempts to provide effective exact solutions for MWT include [16] and [6], who reformulated MWT as an integer linear programming problem and used polyhedral approaches to compute exact solutions. These methods were able to solve larger problem sizes than the DP algorithm in [5], but still do not scale past very modest dataset sizes. Heuristic approaches to MWT have also been developed. Koller and Raidl [7] presented two innovations: a fast heuristic, very similar to Kruskal's algorithm [8], for quickly building an approximate MWT from an alignment graph. They incorporated this heuristic into a genetic algorithm that iteratively refines the highest-scoring solutions. Finally, we have the work by Moreno and Karp [14], where MWT is reformulated as an Implicit Hitting Set Problem, for which the authors present a two-stage approach: a fast greedy heuristic search is used to build initial solutions, which are then refined with integer linear programming. Regrettably, not even the heuristics are able to run on even moderate-sized datasets, and the methods of Moreno and Karp [14] and Koller and Raidl [7] are not publicly available.

To enable MWT approaches to be used on large datasets, we propose the Maximum Weight Trace for Alignment Merger Problem (MWT-AM), a generalization of MWT to the problem of merging a set of disjoint alignments. This extension is motivated by use of divide-and-conquer for large-scale multiple sequence alignment: a set of unaligned sequences is divided into subsets, alignments are computed on each subset (using a preferred method, such as MAFFT [4]) and then the disjoint alignments are treated as constraints and merged together. The success of this divide-and-conquer strategy thus depends on each step, with the third step (how the alignments are merged) previously handled using methods such as OPAL [23], Muscle [2], and combinations of these with transitivity.

Several well known MSA estimation methods use this divide-and-conquer strategy, including SATé [9, 10] and PASTA [11], and also a recently developed method, MAGUS [18]. The specific approach in MAGUS merges disjoint alignments by computing a set of "backbone alignments" that are then used to define weights on pairs of sites from different columns in the constraint alignments. These weights are then used to define an alignment graph (with nodes representing sites in the constraint alignments, and weighted edges representing the information in the backbone alignments), which is then processed to perform an alignment merger using the Graph Clustering Merger (GCM), a technique developed in [18].

Our study, using both biological and simulated datasets with up to 24,000 sequences, shows that the GCM technique used in MAGUS is an effective heuristic for the MWT-AM problem we pose, and some variants of this technique specifically designed to optimize MWT-AM are even more accurate. We call these variants GCM-MWT. We also show that the alignments produced by merging disjoint alignments using GCM-MWT (with weightings produced on pairs of columns using a library graph technique from MAGUS [18]) produce more accurate merged alignments than MAFFT-merge and T-Coffee [15]. Furthermore, using GCM and its variants within the divide-and-conquer pipeline employed by MAGUS enables highly accurate analyses of very large datasets.

See the Supplementary Materials for commands and additional results at https://tandy.cs.illinois.edu/mwt-gcm-alcob-suppl.pdf. All datasets are available online (links provided in Sect. 4).

2   Maximum Weight Trace Alignment Merging

The input to the Maximum Weight Trace (MWT) problem is a set S of sequences and non-negative weights on selected pairs of letters from the sequences; the objective is a multiple sequence alignment of the sequences that optimizes the total weight of the pairs in its columns. The output alignment is also referred to as a "trace", which has a graph-theoretic description in [5] but can be more simply described as follows. A **trace** is a partition of the letters of the input sequences into sets $X_1, X_2, \ldots, X_k$ so that each set has at most one letter from each sequence, and if letters x and y from the same sequence s appear in X_i and X_j respectively with $i < j$, then x appears before y in s. These two properties ensure that the ordering of the sets is consistent with a multiple sequence alignment of the sequences. The **weight** of the trace is the sum of the weights of those pairs that appear in the same subset, and hence in the same column in the alignment defined by the trace.

We generalize the MWT problem to allow the input to be a collection of disjoint alignments instead of a collection of individual sequences, and we refer to this as the Maximum Weight Trace Alignment Merging problem, or MWT-AM. Note that we require that the output alignment induce each of the input alignments, and hence we will refer to these as **constraint alignments**. In this context, a **trace** is a partition of the columns of the input constraint alignments into clusters, which contain at most one column from each alignment and respect a valid ordering: if columns x and y from the same constraint alignment A appear in clusters C_i and C_j, respectively, with $i < j$, then x appears before y in A. Such a trace trivially gives us a valid multiple sequence alignment that induces each of our constraint alignments. Then, given a weighting $w(x, y)$ for every pair of columns x, y from different constraint alignments, we can define the **weight** of a trace T to be $\sum w(x, y)$ over all pairs of columns x, y belonging to the same cluster in T. We now define the MWT-AM problem:

Input: Multiple sequence alignments $A_1, A_2, \ldots, A_k$, with A_i an MSA on set S_i of sequences with $S_i \cap S_j = \emptyset$, and a weight function $w(x, y)$ for every pair of columns x, y with x and y from different MSAs
Output: a trace T on $A_1, A_2, \ldots, A_k$ of maximum weight (i.e., a merged alignment of the input alignments that maximizes the total weight).

It is easy to see that if each MSA A_i is a single sequence, then MWT-AM is identical to MWT.

Theorem 1. *MWT-AM is NP-hard but can be solved in* $O(2^B L^B B^2)$ *for B alignments of total length L.*

The proof is omitted due to space constraints, but follows easily from the corresponding theorems for MWT in [6].

3   Graph Clustering Merger (GCM)

The Graph Clustering Merger (GCM) is an algorithm that was developed for use in MAGUS to merge the constraint alignments it produces in a divide-and-conquer setting. In this original formulation, GCM constructs a set of backbone alignments (i.e., a library) in order to weight the pairs of sites from different constraint alignments, but it can also be used to merge a set of disjoint alignments if the weights are provided by the user. In what follows, we describe GCM for use in both cases, noting that without user-provided weights it will perform an initial step (referred to as Step 0 below) to compute the weights. The Graph Clustering Merger (GCM) then uses the weights it computes (or that it is given) to construct a merged alignment, which is referred to as a "trace" [5], using a sequence of steps. We explore variants of the original GCM algorithm from [18] in an attempt to improve the MWT-AM scores. Hence, the final general design for these GCM variants has several stages (Fig. 1), some of which allow for variants. If the input includes weights on the pairs of sites from different constraint alignments, then GCM skips Step 0. Figure 1 presents a description of the stages of the algorithm.

Variants on Step 1: Alignment Graph. The library graph has a vertex for every site in each of the input alignments, and the weight of the edge between two vertices (i.e., sites) is computed using the default GCM technique as follows. We compute a set of 10 "backbone alignments", each of which contains 200 sequences that are obtained by selecting sequences randomly from the constraint alignments. These sequences are then aligned using existing methods (default: MAFFT -l-ins-i). The weight on a pair of sites from different alignments is the weighted number of pairs of letters (one from each site) that appear in one or more of the backbone alignments (i.e., if x and y are two letters that are aligned in q backbone alignments, then this particular pair (x, y) contributes q to the total weight).

We explored one variant on the alignment graph creation, where we remove edges that appear in the initial alignment graph that go between sites in the same constraint alignment (as these will never appear in the final trace); we refer to this as the "restricted alignment graph".

Variants on Step 2: Clustering. We explore the original technique, which uses the Markov Clustering Algorithm (MCL) from [21], in comparison to two new approaches. Furthermore, within MCL, we vary the inflation factor to evaluate its impact. **Multi-level Regularized MCL (MRL-MCL)** is a modification

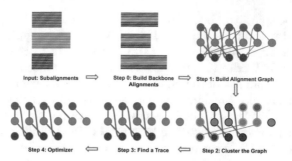

Fig. 1. **The GCM algorithm design.** The input to GCM is a set of constraint alignments. Optionally, we also supply weights on pairs of columns (sites) from different constraint alignments. **Step 0:** If the weights are not provided, this step constructs the backbone alignments, and uses these to compute the weights on pairs of columns. **Step 1:** We build an alignment graph, where each node represents a column from a constraint alignment and the weight of the edge between two nodes is the weight for the corresponding pair of columns. Thicker lines represent edges with higher weight. **Step 2:** We cluster the alignment graph with our method of choice. **Step 3:** We find a valid trace, where each cluster represents a column in our final alignment. Our example shows two common problems: the orange-outlined cluster contains columns from the same constraint alignment, and there is no valid ordering between the green-outlined cluster and the three other clusters that it "crosses". **Step 4:** We optionally run our trace through the optimizer, which will greedily move nodes between clusters to increase the MWT score. (Color figure online)

of MCL [17] to improve scalability and reduce over-clustering. **Region Growing (RG)** is inspired by the heuristic in [7], and is reminiscent of Kruskal's algorithm for finding minimum spanning trees. We initialize our clustering with every node in its own cluster. Each pair of clusters is added to a max heap, weighted by the weight of the edge between them. We then proceed to take pairs of clusters off the heap, merging the pair together if they don't contain any nodes from the same subalignment (i.e. the new cluster would be a valid MSA column). When we merge cluster B into cluster A, we update A's weight to all of B's neighbors (we call clusters neighbors if any of their elements have edges between them); for each cluster C among B's neighbors, we let $Weight(A, C) = Weight(A, C) + Weight(B, C)$ and put the pair (A,C) on the max heap. We continue merging pairs off the heap until we can't merge anything else.

Variants for Step 3: Computing a Trace. We compare the original technique in GCM, called "minclusters", to two new techniques. We explore the **Fiduccia-Mattheyses (FM)** algorithm [3], previously applied to the realm of multiple sequence alignment by MSARC [13]. FM is a heuristic for finding a minimum-weight split in a graph, constrained by a balancing requirement (the sizes of the two parts can differ by at most some value). Like MSARC, we use FM to recursively cut our graph in half, looking for bipartitions that minimize the weight of the split. Notably, the FM trace does not require a prior clustering, meaning that Step 2 can be skipped. The **MWTgreedy** algorithm is extremely similar to the heuristic described in [14], and attempts to resolve a trace with respect to the MWT criterion. In **MWTSEARCH**, we combine MWTgreedy with the lookahead ability of the minclusters search algorithm used in MAGUS. Instead of just greedily breaking the smallest edge in every cycle, we consider every edge in the cycle as a possible move, and we try to find a path of such moves with the smallest weight. The **RG-FAST** method is nearly the same as the Region Growing (RG) clustering method described earlier, but we constrain the algorithm to produce a valid trace, rather than just a clustering.

Variants for Step 4: Optimizer. This is an optional step, which uses a greedy strategy to improve the MWT score, moving nodes between columns in our trace, and stopping when no additional gains can be made. A "move" entails transferring a node X from one cluster (column) to another, in such a way that the trace remains valid. In each iteration, we identify all valid moves (as defined above) with a positive improvement to the MWT score. We perform these moves in descending order of gain, updating the gains of subsequent moves as needed. We stop when we no longer see any gainful moves.

4   Experimental Study

Overview. We explored GCM variants, T-Coffee [15], and MAFFT-merge [4]; to the best of our knowledge, T-Coffee and MAFFT-merge are the only other methods that are designed to merge three or more disjoint alignments. We used simulated and biological datasets from prior studies to explore these methods. For the simulated datasets we know the true alignment, and for the biological datasets we have reference alignments based on structural features.

We evaluated the GCM variants with respect to MWT-AM scores, and then select two variants (one fast and one slower but producing better scores) to further evaluate. We compare these two GCM-MWT variants to T-Coffee and MAFFT-merge with respect to alignment accuracy (computed using FastSP [12]), using the SP-score (i.e., recall, or the fraction of true pairwise homologies recovered in the estimated alignment) and the Modeler Score (i.e., precision, or fraction of the pairwise homologies in the estimated alignment that appear in the true alignment). We also noted wallclock running times. All analyses were run on a single node with 16 CPUs and 64 GB of memory on the Campus Cluster at UIUC. The MAFFT–merge -linsi analyses of the three larger biological datasets (16S.B.ALL, 16S.3, and 16S.T) were run on the tallis cluster, which allows four weeks of analysis and up to 256 GB of memory.

Dataset Generation. All the datasets we explore are nucleotide datasets that are available in online repositories. We used 14 different simulation conditions, each with 20 replicates and containing 1000 sequences, ranging in gap length distribution and difficulty due to rate of evolution. We also analyzed 4 biological datasets with curated alignments based on secondary structure; three of the four biological datasets have at least 5000 sequences and the remaining one has only 740 sequences. In total, we analyzed 284 datasets, and performed multiple analyses on each. We provide the empirical statistics for these datasets in Table S1 of the Supplementary Materials.

Simulated Data. We use 14 model conditions (12 simulated using the ROSE [20] software and used to study PASTA [11] and SATé [9,10]) and 2 subsets of the million-sequence RNASim [11] simulated dataset (called RNASim and RNASim2) that were used in a prior study [19]. Each dataset has 1000 sequences. For the training phase, we use ROSE datasets 1000M1 and 1000M4. We use the remaining model conditions (ROSE datasets 1000L3, 1000S1, 1000M2, 1000L1, 1000S2, 100S3, 1000M3, 10000L2, 1000L4, and 1000S4, and the two RNASim datasets) for the testing phase. The ROSE [20] datasets evolve under *i.i.d.* evolution with substitutions and indels, and under three indel lengths (long, short, and medium); the RNASim and RNASim2 datasets evolve under a model that includes selection and takes RNA secondary structure into consideration, and hence are more complex than the ROSE datasets.

Biological Data. We use three biological datasets from the Comparative Ribosomal Database [1] that have been previously used for benchmarking multiple sequence alignments [10,11,18]; specifically, we used versions of the 16S.M, 16S.3, 16S.T, and 16S.B.ALL datasets, pruned to remove sequences that are more than 20% from the median sequence length (the same datasets studied in [18]).

Creating Disjoint Subset Alignments. Given a sequence dataset, we have two ways of producing the disjoint alignments that are given as input to the methods we study. (1) We randomly decompose the dataset into disjoint subsets or (2) We use PASTA to produce the decomposition (i.e., PASTA computes an initial alignment and tree, and then decomposes the dataset into subsets using the tree by deleting edges). In each case, we decompose the dataset into subsets of approximately equal size. We then compute alignments on each subset using an existing method (default: MAFFT -l-ins-i [4]), thus producing a set of disjoint alignments.

Data Availability. Please see:

– https://doi.org/10.5061/dryad.95x69p8h8 for the RNASim2 dataset
– https://databank.illinois.edu/datasets/IDB-2643961 for the remaining datasets

Experiments. Experiments 0 and 1 use the training datasets (1000M1 and 1000M4) and Experiment 2 uses the remaining datasets (i.e., "testing datasets"). In Experiment 0, we explored the correlation between MWT-AM scores and alignment accuracy. In Experiment 1 (training), we explored variants of GCM to select algorithmic parameters that performed well for the MWT-AM criterion; this produced two variants of GCM, one that is designed for speed on large datasets (GCM-fast) and one that is slower but generally has better MWT-AM scores (GCM-slow). In Experiment 2, we evaluated the best GCM variants (GCM-fast and GCM-slow) on the testing datasets in comparison to T-Coffee and MAFFT–merge.

5   Results

Due to space constraints, see Supplementary Materials for additional details, results, and discussion.

Results for Experiment 0. We explored the correlations between MWT-AM criterion and alignment accuracy using the default version of GCM on 1000M1 and 1000M4, and under two ways of decomposing the dataset: random decompositions and decompositions obtained during the first iteration of a PASTA alignment. All correlations were at least moderately strong, with higher correlations on the datasets that were more challenging to align (Supplementary Materials, Table S2 and Fig. S1).

Results for Experiment 1. The default settings for each step in GCM was as good or better than nearly all the alternatives with respect to MWT-AM score (see Supplementary Materials Tables S3–S5 and Fig. S2). However, a few combinations improved on the default combination implemented in GCM (step 2: MCL, step 3: MINCLUSTERS, step 4: nothing), although each only provided a small improvement. Second, some combinations showed much less accurate results than default mode (e.g., RG+MWTGREEDY or RG by itself at step 3), but all combinations are more or less even in terms of average accuracy when step 4 (the optimizer) is added. Last, the optimizer clearly improves the MWT-AM score. Based on these results, we selected two variants for GCM to explore on the remaining datasets:

- `MCL+MINCLUSTERS` (i.e., no optimizer step at the end); note that this is the default setting used in MAGUS [18], and so we also refer to this as GCM (default).
- `FM+OPTIMIZER` (i.e., no Step 2 where the initial clustering is performed); we refer to this as GCM (fm-opt).

Results for Experiment 2. We compare the GCM variants to MAFFT–merge (two variants) and T-Coffee with respect to alignment accuracy (Modeler Score and SP-score) and running time. For GCM, the weights are computed from the backbone alignments it computes. In these experiments (Supplementary Materials, Fig. S3), T-Coffee is much less accurate than the other methods. Based on a discussion with the developer, Cedric Notredame, T-Coffee is not designed for datasets of this size and has also not been tested sufficiently on nucleotide datasets. Hence, the use of T-Coffee for merging large nucleotide alignments may not be appropriate. The rest of this section focuses on the comparison between the GCM variants and MAFFT–merge.

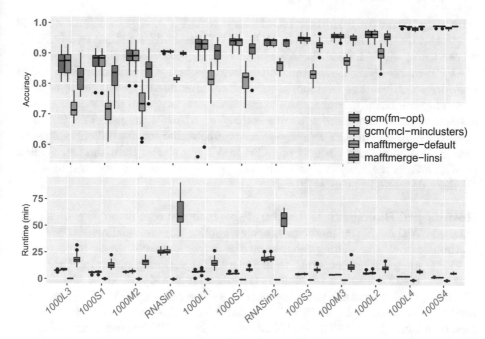

Fig. 2. Experiment 2: Results on the simulated datasets. Accuracy refers to the average of SP-score (i.e., recall) and Modeler score (i.e., precision) in the estimated alignments. Running time is based on wallclock time, using the Campus Cluster at the University of Illinois constrained to machines with the same amount of memory. All analyses here were performed on datasets produced in the first iteration of a default PASTA analysis.

Accuracy on Simulated Datasets. Figure 2 shows the following trends. The two ways of running MAFFT–merge are clearly distinguished, with the L-ins-i version much more accurate than the default version. Both ways of running GCM-MWT are approximately equal in accuracy, and GCM-MWT is almost always more accurate than MAFFT–merge using L-ins-i, with the only exceptions being the easiest datasets (1000L4 and 1000S4) that have the lowest rate of evolution. Analyses using random decompositions show the same trends, and are provided in the supplementary materials (Fig. S4).

One additional observation is that MAFFT–merge using -linsi performs relatively well, although we used it in a way that is not recommended on the MAFFT–merge website, which recommends that MAFFT–merge use monophyletic clusters as entries (https://mafft.cbrc.jp/alignment/software/merge.html). Although monophyly is not ensured in PASTA decompositions, it is definitely violated in random decompositions, showing that MAFFT–merge (using -linsi) is a valuable approach for merging alignments of non-monophyletic clusters, even though it wasn't designed for this case.

Running Time on Simulated Datasets: MAFFT–merge in default mode is the fastest method, followed by the GCM-MWT methods, and then by MAFFT–merge using -linsi. Indeed, both ways of running GCM-MWT are faster than MAFFT–merge using L-ins-i, and often twice as fast. Overall, therefore, GCM provides an advantage over both ways of running MAFFT–merge: it is more accurate than both (except for the easiest datasets, where all methods are good) and faster than the more accurate setting of MAFFT–merge. The improvement of GCM over MAFFT–merge is most substantial for the difficult datasets. Finally, although both variants of GCM-MWT are slower than the fast version of MAFFT–merge, they are faster than the more accurate version, and finish on all these datasets in at most 30 min.

Results on Biological Datasets: Table 1 shows results on the biological datasets. Although the methods are very similar on the smallest dataset, the methods are clearly distinguishable on the larger datasets. Most importantly, MAFFT–merge is less accurate than both GCM variants, obtaining SP-scores that are 10–13% lower and Modeler scores that are 4–6% lower than GCM (default). It is important to realize that the -linsi option for MAFFT–merge is unable to complete in 4 h on these larger datasets, and so we ran MAFFT–merge-linsi on a tallis node (16 CPUs, 256 GB memory) for 50 h on the large biological datasets (16S.3, 16S.T and 16S.B.ALL). A comparison between the two GCM variants shows that they have the same average alignment accuracy (where we average SP-score and Modeler score), but GCM (default) has somewhat better Modeler score and GCM (fm+opt) has somewhat better SP-score.

Table 1. Experiment 2: Results on the biological datasets; the best results are bold-faced (ties are for methods within 1% of the best found result)

Dataset (# seqs)	Method	SP-Score	Modeler	Avg.	Time (min.)
16S.M (740)	GCM (default)	**88.4**	**86.9**	**87.7**	5.8
	GCM (fm+opt)	**88.4**	**86.7**	**87.6**	5.9
	MAFFTmerge (l-ins-i)	**88.5**	**87.0**	**87.7**	6.0
	MAFFTmerge (default)	85.9	86.7	86.3	0.02
16S.3 (5,489)	GCM (default)	**92.1**	**86.5**	**89.3**	12.4
	GCM (fm+opt)	**92.8**	**85.9**	**89.3**	31.6
	MAFFTmerge (l-ins-i)	91.6	**85.8**	88.8	2926.5
	MAFFTmerge (default)	81.7	81.4	81.5	0.5
16S.T (5,548)	GCM (default)	**92.4**	**87.7**	**90.1**	15.0
	GCM (fm+opt)	**92.7**	86.4	89.5	37.7
	MAFFTmerge (l-ins-i)	91.4	86.2	88.8	2607.1
	MAFFTmerge (default)	81.7	82.3	82.0	0.6
16S.B.ALL (24,246)	GCM (default)	**95.8**	**95.7**	**95.7**	13.8
	MAFFTmerge (default)	83.4	90.5	87.0	7.0

A comparison of running times shows the following trends. On the small dataset we used MAFFT–merge with the –linsi setting, which is more accurate but also more expensive; as a result, the three methods have nearly identical running times (6 min). On the three larger datasets, MAFFT–merge is much faster, finishing in under a minute on 16S.3 and 16S.T (each with about 5500 sequences), and in 7 min on the 16S.B.ALL dataset (with 24,246 sequences); in contrast, the two GCM variants take longer. As expected, GCM (default) is faster, with 12–15 min on all three datasets, but GCM (fm+opt) is slower: 31–38 min on 16S.3 and 16S.T, and then unable to complete on the 16S.B.ALL dataset within the allowed time.

6   Conclusions

Here we introduced the MWT-AM (MWT for Alignment Merging) problem. Our study shows that GCM-MWT (default) and GCM-MWT (fm+opt), two variants of the Graph Clustering Merger (GCM) method introduced in [18], provide good accuracy for the MWT-AM problem (with a slight advantage to GCM-MWT (fm+opt)). A comparison between the two GCM-MWT variants shows that both provide about the same alignment accuracy (where we average the Modeler and SP-scores), with a slight advantage in SP-score to GCM-MWT (fm+opt) and a slight advantage in Modeler score to GCM-MWT (default). However, GCM-MWT (default) is faster than GCM-MWT (fm+opt), and so can be used on larger datasets. Thus, the choice between the two methods may depend mainly on whether recall (SP-score) or precision (Modeler score) is more important, but dataset size and available computational resources will also be a consideration.

We also showed that using GCM-MWT improved on T-Coffee and MAFFT–merge with respect to alignment accuracy when merging disjoint alignments. Future work should examine other techniques (e.g., M-Coffee method [22]) for merging three or more disjoint alignments. To the best of our knowledge, there are no results regarding polynomial-time constant-factor approximations for MWT and MWT-AM; future work should investigate this. The scalability of GCM to very large datasets has not yet been explored, and so this is an additional direction for future research. Finally, although our focus has been on using GCM-MWT for nucleotide alignments, this approach might be useful for protein alignments, in which case T-Coffee might provide an advantage.

Acknowledgments. This work was supported in part by NSF ABI-1458652 to TW and by the Debra and Ira Cohen fellowship to VS.

References

1. Cannone, J.J., et al.: The comparative RNA Web (CRW) site: an online database of comparative sequence and structure information for ribosomal, intron, and other RNAs. BMC Bioinf. **3**(1), 1–31 (2002). https://doi.org/10.1186/1471-2105-3-2

2. Edgar, R.C.: MUSCLE: a multiple sequence alignment method with reduced time and space complexity. BMC Bioinf. **5**(1), 113 (2004)

3. Fiduccia, C.M., Mattheyses, R.M.: A linear-time heuristic for improving network partitions. In: 19th Design Automation Conference, pp. 175–181. IEEE (1982)

4. Katoh, K., Kuma, K.I., Toh, H., Miyata, T.: MAFFT version 5: improvement in accuracy of multiple sequence alignment. Nucleic Acids Res. **33**(2), 511–518 (2005)

5. Kececioglu, J.: The maximum weight trace problem in multiple sequence alignment. In: Apostolico, A., Crochemore, M., Galil, Z., Manber, U. (eds.) CPM 1993. LNCS, vol. 684, pp. 106–119. Springer, Heidelberg (1993). https://doi.org/10.1007/BFb0029800

6. Kececioglu, J.D., Lenhof, H.P., Mehlhorn, K., Mutzel, P., Reinert, K., Vingron, M.: A polyhedral approach to sequence alignment problems. Discrete Appl. Math. **104**(1–3), 143–186 (2000)

7. Koller, G., Raidl, G.R.: An evolutionary algorithm for the maximum weight trace formulation of the multiple sequence alignment problem. In: Yao, X., et al. (eds.) PPSN 2004. LNCS, vol. 3242, pp. 302–311. Springer, Heidelberg (2004). https://doi.org/10.1007/978-3-540-30217-9_31

8. Kruskal, J.B.: On the shortest spanning subtree of a graph and the traveling salesman problem. Proc. Am. Math. Soc. **7**(1), 48–50 (1956)

9. Liu, K., Raghavan, S., Nelesen, S., Linder, C.R., Warnow, T.: Rapid and accurate large-scale coestimation of sequence alignments and phylogenetic trees. Science **324**(5934), 1561–1564 (2009)

10. Liu, K., et al.: SATé-II: very fast and accurate simultaneous estimation of multiple sequence alignments and phylogenetic trees. Syst. Biol. **61**(1), 90 (2012)

11. Mirarab, S., Nguyen, N., Guo, S., Wang, L.S., Kim, J., Warnow, T.: PASTA: ultra-large multiple sequence alignment for nucleotide and amino-acid sequences. J. Comput. Biol. **22**(5), 377–386 (2015)

12. Mirarab, S., Warnow, T.: FASTSP: linear time calculation of alignment accuracy. Bioinformatics **27**(23), 3250–3258 (2011)

13. Modzelewski, M., Dojer, N.: MSARC: multiple sequence alignment by residue clustering. Alg. Mol. Biol. **9**(1), 12 (2014)

14. Moreno-Centeno, E., Karp, R.M.: The implicit hitting set approach to solve combinatorial optimization problems with an application to multigenome alignment. Oper. Res. **61**(2), 453–468 (2013)

15. Notredame, C., Higgins, D.G., Heringa, J.: T-Coffee: a novel method for fast and accurate multiple sequence alignment. J. Mol. Biol. **302**(1), 205–217 (2000)

16. Reinert, K., Lenhof, H.P., Mutzel, P., Mehlhorn, K., Kececioglu, J.D.: A branch-and-cut algorithm for multiple sequence alignment. In: Proceedings of the First Annual International Conference on Computational Molecular Biology (RECOMB), pp. 241–250 (1997)

17. Satuluri, V., Parthasarathy, S.: Scalable graph clustering using stochastic flows: applications to community discovery. In: Proceedings of the 15th ACM SIGKDD International Conference on Knowledge Discovery and Data Mining, pp. 737–746 (2009)

18. Smirnov, V., Warnow, T.: MAGUS: multiple sequence alignment using graph clustering. Bioinformatics (2020)

19. Smirnov, V., Warnow, T.: Phylogeny estimation given sequence length heterogeneity. Syst. Biol. **70**(2), 268–282 (2020)

20. Stoye, J., Evers, D., Meyer, F.: Rose: generating sequence families. Bioinformatics **14**(2), 157–163 (1998)

21. Van Dongen, S.M.: A cluster algorithm for graphs. Technical report, National Research Institute for Mathematics and Computer Science in the Netherlands, Amsterdam, iNS-R0010, May 2000
22. Wallace, I.M., O'sullivan, O., Higgins, D.G., Notredame, C.: M-Coffee: combining multiple sequence alignment methods with T-Coffee. Nucleic Acids Res. **34**(6), 1692–1699 (2006)
23. Wheeler, T.J., Kececioglu, J.D.: Multiple alignment by aligning alignments. Bioinformatics **23**(13), i559–i568 (2007)

Author Index

Printed in the United States
by Baker & Taylor Publisher Services